HEINEMANN MODULAR MATHEMATICS
for
EDEXCEL AS AND A-LEVEL

Mechanics 5

John Hebborn Jean Littlewood

1

2

3

4

Heinemann

Edexcel
Success through qualifications

Heinemann Educational Publishers,
a division of Heinemann Publishers (Oxford) Ltd,
Halley Court, Jordan Hill, Oxford, OX2 8EJ

OXFORD MELBOURNE AUCKLAND JOHANNESBUURG
BLANTYRE GABORONE PORTSMOUTH NH (USA) CHICAGO

First published 2001

05 04 03 02
10 9 8 7 6 5 4 3 2

ISBN 0 435 51078 9

Cover design by Gecko Limited

Original design by Geoffrey Wadsley: additional design work by Jim Turner

Typeset and illustrated by Tech-Set Limited, Gateshead, Tyne and Wear

Printed in Great Britain by Scotprint

Acknowledgements:

The publisher's and authors' thanks are due to Edexcel for permission to reproduce questions from past examination papers. These are marked with an [E].

The answers have been provided by the authors and are not the responsibility of the examining board.

About this book

This book is designed to provide you with the best preparation possible for your Edexcel M5 exam. The series authors are senior examiners and exam moderators themselves and have a good understanding of Edexcel's requirements.

Use this **new edition** to prepare for the new 6-unit specification. Use the first edition (*Heinemann Modular Mathematics for London AS and A-Level*) if you are preparing for the 4-module syllabus.

Finding your way around

To help to find your way around when you are studying and revising use the:

- edge marks (shown on the front page) – these help you to get to the right chapter quickly;
- contents list – this lists the headings that identify key syllabus ideas covered in the book so you can turn straight to them;
- index – if you need to find a topic the bold number shows where to find the main entry on a topic.

Remembering key ideas

We have provided clear explanations of the key ideas and techniques you need throughout the book. Key ideas you need to remember are listed in a summary of key points at the end of each chapter and marked like this in the chapters:

■ **K.E. of body** $= \frac{1}{2}I\omega^2$

Exercises and exam questions

In this book questions are carefully graded so they increase in difficulty and gradually bring you up to exam standard.

- past exam questions are marked with an [E];
- review exercises on pages 39 and 119 help you practise answering questions from several areas of mathematics at once, as in the real exam;
- exam style practice paper – this is designed to help you prepare for the exam itself;
- answers are included at the end of the book – use them to check your work.

Contents

Applications of vectors in mechanics

1

1.1 Solution of simple vector differential equations

In Book M3 and Book M4, in the study of the motion of a particle moving in a straight line, several differential equations were encountered. The solutions of differential equations of the form $\frac{\mathrm{d}v}{\mathrm{d}t} = \mathrm{f}(v)$, $\frac{\mathrm{d}^2x}{\mathrm{d}t^2} + \omega^2 x = 0$ (simple harmonic motion) and $\frac{\mathrm{d}^2x}{\mathrm{d}t^2} + \frac{\mathrm{d}x}{\mathrm{d}t} + \omega^2 x = 0$ (damped harmonic motion) were considered. When a particle is moving in a plane, or in three-dimensional space, the use of vectors can facilitate the study of the motion. The resulting differential equations will then involve vectors.

Motion in a plane when the acceleration is proportional to the velocity

The equation of motion for this type of motion is of the form

$$\frac{\mathrm{d}\mathbf{v}}{\mathrm{d}t} = k\mathbf{v}$$

This equation appears to be an equation of the 'variables separable' type. However, such an approach suggests the equation

$$\int \frac{\mathrm{d}\mathbf{v}}{\mathbf{v}} = \int \mathrm{d}t$$

But the left-hand side of this equation involves both division by a vector and integration with respect to the vector $\mathbf{v}$. Neither of these operations is defined. In order to avoid these problems, start by substituting:

$$\mathbf{v} = u\mathbf{i} + w\mathbf{j}$$

Then $$\frac{\mathrm{d}\mathbf{v}}{\mathrm{d}t} = \frac{\mathrm{d}u}{\mathrm{d}t}\mathbf{i} + \frac{\mathrm{d}w}{\mathrm{d}t}\mathbf{j}$$

and the equation $\frac{d\mathbf{v}}{dt} = k\mathbf{v}$ becomes

$$\frac{du}{dt}\mathbf{i} + \frac{dw}{dt}\mathbf{j} = k(u\mathbf{i} + w\mathbf{j})$$

Equating coefficients of **i** and **j** on the two sides of this equation gives

$$\frac{du}{dt} = ku \quad \text{and} \quad \frac{dw}{dt} = kw$$

Each of these equations can be solved by separating the variables:

$$\int \frac{du}{u} = k\int dt$$

$$\ln|u| = kt + \ln a$$

where a is a constant

$$\ln\left(\frac{|u|}{a}\right) = kt$$

$$\frac{u}{a} = e^{kt}$$

$$u = ae^{kt}$$

$$\int \frac{dw}{w} = k\int dt$$

$$\ln|w| = kt + \ln b$$

where b is a constant

$$\ln\left(\frac{|w|}{b}\right) = kt$$

$$\frac{w}{b} = e^{kt}$$

$$w = be^{kt}$$

Hence $\quad \mathbf{v} = u\mathbf{i} + w\mathbf{j}$

$$\mathbf{v} = ae^{kt}\mathbf{i} + be^{kt}\mathbf{j}$$

$$\mathbf{v} = e^{kt}(a\mathbf{i} + b\mathbf{j})$$

and so $\quad \mathbf{v} = \mathbf{k}e^{kt}$

where **k** is a constant vector.

Example 1

At time t seconds, the velocity $\mathbf{v}\,\text{m s}^{-1}$ of a particle P, moving in a plane, satisfies the differential equation

$$\frac{d\mathbf{v}}{dt} = 2\mathbf{v}$$

Given that when $t = 0$, the position vector of P, **r** metres, relative to a fixed origin, is $(\mathbf{i} - \mathbf{j})$ m and $\mathbf{v} = (2\mathbf{i} + 3\mathbf{j})$, find an expression for **r** in terms of t.

$$\frac{d\mathbf{v}}{dt} = 2\mathbf{v}$$

Substituting $\quad \mathbf{v} = u\mathbf{i} + w\mathbf{j}$

gives $\quad \frac{du}{dt}\mathbf{i} + \frac{dw}{dt}\mathbf{j} = 2(u\mathbf{i} + w\mathbf{j})$

and so $\frac{du}{dt} = 2u \qquad \frac{dw}{dt} = 2w$

Separating the variables:

$$\int \frac{du}{u} = \int 2dt \qquad\qquad \int \frac{dw}{w} = \int 2dt$$

$$\ln|u| = 2t + a \qquad\qquad \ln|w| = 2t + b$$

When $t = 0$, $\mathbf{v} = 2\mathbf{i} + 3\mathbf{j}$ so $u = 2$ and $w = 3$. Hence $\ln 2 = a$ and $\ln 3 = b$.

So $\ln|u| = 2t + \ln 2 \qquad\qquad \ln|w| = 2t + \ln 3$

$$\ln\left(\frac{|u|}{2}\right) = 2t \qquad\qquad \ln\left(\frac{|w|}{3}\right) = 2t$$

$$u = 2e^{2t} \qquad\qquad w = 3e^{2t}$$

So $\mathbf{v} = 2e^{2t}\mathbf{i} + 3e^{2t}\mathbf{j}$

and $\frac{d\mathbf{r}}{dt} = 2e^{2t}\mathbf{i} + 3e^{2t}\mathbf{j}$

This first-order differential equation poses no problems, as the solution involves integration with respect to the *scalar* quantity t and **i** and **j** are constant vectors.

Integrating with respect to t gives

$$\mathbf{r} = \int (2e^{2t}\mathbf{i} + 3e^{2t}\mathbf{j})\, dt$$

$$\mathbf{r} = e^{2t}\mathbf{i} + \tfrac{3}{2}e^{2t}\mathbf{j} + c\mathbf{i} + d\mathbf{j}$$

where c and d are constants.

As $\mathbf{r} = \mathbf{i} - \mathbf{j}$ when $t = 0$,

$$\mathbf{i} - \mathbf{j} = \mathbf{i} + \tfrac{3}{2}\mathbf{j} + c\mathbf{i} + d\mathbf{j}$$

Equating coefficients of **i** on the two sides of this equation gives

$$1 = 1 + c$$

so $c = 0$

Equating coefficients of **j** on the two sides of this equation gives

$$-1 = \tfrac{3}{2} + d$$

so $d = -\tfrac{5}{2}$

Therefore $\mathbf{r} = e^{2t}\mathbf{i} + (\tfrac{3}{2}e^{2t} - \tfrac{5}{2})\mathbf{j}$

Motion in three-dimensional space when the acceleration is proportional to the velocity

The equation of motion for this type of motion is the same as when the movement is restricted to a plane. That is, the equation of motion is

$$\frac{d\mathbf{v}}{dt} = k\mathbf{v}$$

However, as we are now dealing with three-dimensional vectors, the required substitution is

$$\mathbf{v} = u\mathbf{i} + w\mathbf{j} + s\mathbf{k}$$

Comparing the three-dimensional and two-dimensional substitutions and considering the two-dimensional solution shows that the three-dimensional solution now becomes

$$\mathbf{v} = e^{kt}(a\mathbf{i} + b\mathbf{j} + c\mathbf{k})$$

or $$\mathbf{v} = \mathbf{K}e^{kt}$$

where **K** is a constant three-dimensional vector.

Solution of the vector form of the equation for damped harmonic motion

In order to solve the equation

$$\frac{d^2\mathbf{r}}{dt^2} + 2k\frac{d\mathbf{r}}{dt} + (k^2 + n^2)\mathbf{r} = \mathbf{0}$$

substitute $\mathbf{r} = x\mathbf{i} + y\mathbf{j}$ to obtain scalar equations, as before.

The equation then becomes

$$\left(\frac{d^2x}{dt^2}\mathbf{i} + \frac{d^2y}{dt^2}\mathbf{j}\right) + 2k\left(\frac{dx}{dt}\mathbf{i} + \frac{dy}{dt}\mathbf{j}\right) + (k^2 + n^2)(x\mathbf{i} + y\mathbf{j}) = \mathbf{0}$$

This gives the two scalar equations

$$\frac{d^2x}{dt^2} + 2k\frac{dx}{dt} + (k^2 + n^2)x = 0 \qquad (1)$$

and $$\frac{d^2y}{dt^2} + 2k\frac{dy}{dt} + (k^2 + n^2)y = 0 \qquad (2)$$

These equations can be solved using the methods described in Book P4, chapter 6.

The auxiliary equation for equation (1) is

$$\lambda^2 + 2k\lambda^2 + (k^2 + n^2) = 0$$

(The auxiliary equation for equation (2) is the same.)

This may be written as

$$(\lambda + k)^2 + n^2 = 0$$

or $$(\lambda + k)^2 = -n^2$$

So $$\lambda + k = \pm n\mathrm{i}$$

and: $$\lambda = -k \pm n\mathrm{i}$$

(You can also obtain this result by using the quadratic formula to solve the equation.)

Hence the general solutions of equations (1) and (2) are

$$x = \mathrm{e}^{-kt}(A\cos nt + B\sin nt)$$

and $$y = \mathrm{e}^{-kt}(\alpha\cos nt + \beta\sin nt)$$

where A, B, α and β are constants.

And so $\quad \mathbf{r} = \mathrm{e}^{-kt}[(A\cos nt + B\sin nt)\mathbf{i} + (\alpha\cos nt + \beta\sin nt)\mathbf{j}]$

or $\quad \mathbf{r} = \mathrm{e}^{-kt}[(A\mathbf{i} + \alpha\mathbf{j})\cos nt + (B\mathbf{i} + \beta\mathbf{j})\sin nt]$

Writing $\quad \mathbf{a} = A\mathbf{i} + \alpha\mathbf{j}$

and $\quad \mathbf{b} = B\mathbf{i} + \beta\mathbf{j}$

gives $\quad \mathbf{r} = \mathrm{e}^{-kt}(\mathbf{a}\cos nt + \mathbf{b}\sin nt)$

where **a** and **b** are constant vectors.

Notice that this has the same form as the general solution of the differential equation

$$\frac{\mathrm{d}^2z}{\mathrm{d}t^2} + 2k\frac{\mathrm{d}z}{\mathrm{d}t} + (k^2 + n^2)z = 0$$

with the arbitrary scalar constants being replaced by arbitrary constant vectors.

The above solution was obtained assuming that the motion took place in a plane, as the substitution used was $\mathbf{r} = x\mathbf{i} + y\mathbf{j}$. The arbitrary constant vectors in the solution in this case are two-dimensional. For three-dimensional motion the substitution required is $\mathbf{r} = x\mathbf{i} + y\mathbf{j} + z\mathbf{k}$, and the arbitrary constant vectors in the resulting solution will be three-dimensional.

Example 2

The position vector **r** metres, relative to a fixed origin O, of the particle P at time t seconds is such that **r** satisfies the differential equation

$$\frac{\mathrm{d}^2\mathbf{r}}{\mathrm{d}t^2} - 2\frac{\mathrm{d}\mathbf{r}}{\mathrm{d}t} + 5\mathbf{r} = \mathbf{0}$$

Given that when $t = 0$, $\mathbf{r} = 2\mathbf{i} - \mathbf{j}$ and $\dfrac{d\mathbf{r}}{dt} = \mathbf{i} + \mathbf{j}$, find

(a) an expression for $\mathbf{r}$ in terms of t,
(b) the distance of the particle from O when $t = 1$.

(a) $$\frac{d^2\mathbf{r}}{dt^2} - 2\frac{d\mathbf{r}}{dt} + 5\mathbf{r} = \mathbf{0}$$

The auxiliary equation is

$$\lambda^2 - 2\lambda + 5 = 0$$
$$(\lambda - 1)^2 - 1 + 5 = 0$$
$$(\lambda - 1)^2 = -4$$
$$\lambda = 1 \pm 2\mathrm{i}$$

Hence the general solution is

$$\mathbf{r} = e^t(\mathbf{a}\cos 2t + \mathbf{b}\sin 2t) \qquad (1)$$

Since $\mathbf{r} = 2\mathbf{i} - \mathbf{j}$ when $t = 0$:

$$2\mathbf{i} - \mathbf{j} = \mathbf{a} \qquad (2)$$

as $e^0 = 1$, $\cos 0 = 1$ and $\sin 0 = 0$.

Differentiating equation (1) with respect to t gives (by the product rule)

$$\frac{d\mathbf{r}}{dt} = e^t(\mathbf{a}\cos 2t + \mathbf{b}\sin 2t) + e^t(-2\mathbf{a}\sin 2t + 2\mathbf{b}\cos 2t)$$

Since $\dfrac{d\mathbf{r}}{dt} = \mathbf{i} + \mathbf{j}$ when $t = 0$,

$$\mathbf{i} + \mathbf{j} = \mathbf{a} + 2\mathbf{b}$$

From equation (2), $\mathbf{a} = 2\mathbf{i} - \mathbf{j}$, so

$$\mathbf{i} + \mathbf{j} = 2\mathbf{i} - \mathbf{j} + 2\mathbf{b}$$
$$2\mathbf{b} = -\mathbf{i} + 2\mathbf{j}$$

so $$\mathbf{b} = -\tfrac{1}{2}\mathbf{i} + \mathbf{j}$$

and $$\mathbf{r} = e^t[(2\mathbf{i} - \mathbf{j})\cos 2t + (-\tfrac{1}{2}\mathbf{i} + \mathbf{j})\sin 2t]$$

(b) When $t = 1$, $\mathbf{r} = e[(2\mathbf{i} - \mathbf{j})\cos 2 + (-\tfrac{1}{2}\mathbf{i} + \mathbf{j})\sin 2]$
$= e[(2\cos 2 - \tfrac{1}{2}\sin 2)\mathbf{i} + (\sin 2 - \cos 2)\mathbf{j}]$

The distance of the particle from O is $|\mathbf{r}|$ where

$$|\mathbf{r}| = e\sqrt{[(2\cos 2 - \tfrac{1}{2}\sin 2)^2 + (\sin 2 - \cos 2)^2]}$$

Remember that radians must be used for this calculation.

So $$|\mathbf{r}| = 5.02$$

The distance of the particle from O when $t = 1$ is 5.02 m.

Solution of the vector equation $\dfrac{d^2\mathbf{r}}{dt^2} + 2k\dfrac{d\mathbf{r}}{dt} + (k^2 + n^2)\mathbf{r} = \mathbf{g}(t)$ where $\mathbf{g}(t)$ is a vector which depends on t

In order to find a general solution for this type of vector differential equation you must first solve the equation

$$\frac{d^2\mathbf{r}}{dt^2} + 2k\frac{d\mathbf{r}}{dt} + (k^2 + n^2)\mathbf{r} = \mathbf{0}$$

This is the equation for damped harmonic motion considered above. The solution of this equation is called the **complementary function** (see Book P4, chapter 6). Then you must find a **particular integral** for the equation

$$\frac{d^2\mathbf{r}}{dt^2} + 2k\frac{d\mathbf{r}}{dt} + (k^2 + n^2)\mathbf{r} = \mathbf{g}(t)$$

A particular integral is found in the same way as for ordinary differential equations (Book P4, chapter 6), that is, a suitable form with constants is chosen. For a vector equation, the constants are replaced by constant vectors. This is then differentiated and the resulting differentials substituted in the equation to obtain the particular values of the constants. As in Book P4, chapter 6, the complete solution can then be obtained:

- **The general solution of the differential equation**

  $$\frac{d^2\mathbf{r}}{dt^2} + 2k\frac{d\mathbf{r}}{dt} + (k^2 + n^2)\mathbf{r} = \mathbf{g}(t) \text{ is}$$

  $\mathbf{r} =$ complementary function + particular integral

Example 3

Obtain the general solution of the vector differential equation

$$\frac{d^2\mathbf{r}}{dt^2} - 5\frac{d\mathbf{r}}{dt} + 6\mathbf{r} = e^{-t}(\mathbf{i} + \mathbf{j})$$

The complementary function is the general solution of

$$\frac{d^2\mathbf{r}}{dt^2} - 5\frac{d\mathbf{r}}{dt} + 6\mathbf{r} = 0$$

The auxilary equation is

$$\lambda^2 - 5\lambda + 6 = 0$$
$$(\lambda - 3)(\lambda - 2) = 0$$
$$\lambda = 3 \text{ or } \lambda = 2$$

So the complementary function is $\mathbf{r} = \mathbf{a}e^{3t} + \mathbf{b}e^{2t}$ where **a** and **b** are arbitrary vectors. For a particular integral try $\mathbf{r}_p = \mathbf{A}e^{-t}$ where **A** is a constant vector which can be found by substituting $\mathbf{r}_p$ into the differential equation.

As $$\mathbf{r}_p = \mathbf{A}e^{-t}$$

then $$\frac{d\mathbf{r}_p}{dt} = -\mathbf{A}e^{-t}$$

and $$\frac{d^2\mathbf{r}_p}{dt^2} = \mathbf{A}e^{-t}$$

So $$e^{-t}(\mathbf{A} - (-5)\mathbf{A} + 6\mathbf{A}) = e^{-t}(\mathbf{i} + \mathbf{j})$$

$$12\mathbf{A} = \mathbf{i} + \mathbf{j}$$

$$\mathbf{A} = \tfrac{1}{12}(\mathbf{i} + \mathbf{j})$$

Hence the general solution is $\mathbf{r} = \mathbf{a}e^{3t} + \mathbf{b}e^{2t} + \frac{1}{12}(\mathbf{i} + \mathbf{j})e^{-t}$.

Example 4

Find the general solution of the vector differential equation

$$\frac{d^2\mathbf{r}}{dt^2} - 2\frac{d\mathbf{r}}{dt} + 5\mathbf{r} = 10\sin t\,\mathbf{i}$$

The complementary function for this equation was obtained in example 2. It is $\mathbf{r} = e^{t}(\mathbf{a}\cos 2t + \mathbf{b}\sin 2t)$, where **a** and **b** are arbitrary constants. For a particular integral try $\mathbf{r}_p = \mathbf{A}\sin t + \mathbf{B}\cos t$ where **A** and **B** are constant vectors.

Then $$\frac{d\mathbf{r}_p}{dt} = \mathbf{A}\cos t - \mathbf{B}\sin t$$

and $$\frac{d^2\mathbf{r}_p}{dt^2} = -\mathbf{A}\sin t - \mathbf{B}\cos t$$

Substituting these in the differential equation gives

$$-\mathbf{A}\sin t - \mathbf{B}\cos t - 2(\mathbf{A}\cos t - \mathbf{B}\sin t) + 5(\mathbf{A}\sin t + \mathbf{B}\cos t)$$
$$= 10\sin t\,\mathbf{i}$$

Equating terms in $\sin t$ gives

$$-\mathbf{A} + 2\mathbf{B} + 5\mathbf{A} = 10\mathbf{i}$$

or $$4\mathbf{A} + 2\mathbf{B} = 3\mathbf{i} \qquad (1)$$

Equating terms in $\cos t$ gives

$$-\mathbf{B} - 2\mathbf{A} + 5\mathbf{B} = \mathbf{0}$$

or $$-\mathbf{A} + 2\mathbf{B} = \mathbf{0} \qquad (2)$$

Equation (1) – equation (2) gives

$$5\mathbf{A} = 10\mathbf{i}$$

$$\mathbf{A} = 2\mathbf{i}$$

Substituting in equation (2) gives

$$\mathbf{B} = \mathbf{i}$$

So a particular integral is $\mathbf{r}_\mathrm{p} = 2\mathbf{i}\sin t + \mathbf{i}\cos t$ and the general solution is $\mathbf{r} = \mathrm{e}^t(\mathbf{a}\cos 2t + \mathbf{b}\sin 2t) + 2\mathbf{i}\sin t + \mathbf{i}\cos t$.

Example 5

The position vector $\mathbf{r}$ metres, relative to a fixed origin O, of the particle P at time t seconds is such that $\mathbf{r}$ satisfies the differential equation

$$\frac{\mathrm{d}^2\mathbf{r}}{\mathrm{d}t^2} - 2\frac{\mathrm{d}\mathbf{r}}{\mathrm{d}t} + 5\mathbf{r} = 10\sin t\,\mathbf{i}$$

Given that when $t = 0$, $\mathbf{r} = 2\mathbf{i} - \mathbf{j}$ and $\frac{\mathrm{d}\mathbf{r}}{\mathrm{d}t} = \mathbf{i} + \mathbf{j}$ find an expression for $\mathbf{r}$ in terms of t.

The general solution for the equation was found in example 4:

$$\mathbf{r} = \mathrm{e}^t(\mathbf{a}\cos 2t + \mathbf{b}\sin 2t) + 2\mathbf{i}\sin t + \mathbf{i}\cos t \qquad (1)$$

In order to determine the values for the constants $\mathbf{a}$ and $\mathbf{b}$, the given initial conditions must be used. Since $\mathbf{r} = 2\mathbf{i} - \mathbf{j}$ when $t = 0$,

$$2\mathbf{i} - \mathbf{j} = \mathbf{a} + \mathbf{i}$$

$$\mathbf{a} = \mathbf{i} - \mathbf{j} \qquad (2)$$

Differentiating equation (1) with respect to t gives

$$\frac{\mathrm{d}\mathbf{r}}{\mathrm{d}t} = \mathrm{e}^t(\mathbf{a}\cos 2t + \mathbf{b}\sin 2t) + \mathrm{e}^t(-2\mathbf{a}\sin 2t + 2\mathbf{b}\cos 2t)$$
$$+ 2\mathbf{i}\cos t - \mathbf{i}\sin t$$

Since $\frac{\mathrm{d}\mathbf{r}}{\mathrm{d}t} = \mathbf{i} + \mathbf{j}$ when $t = 0$:

$$\mathbf{i} + \mathbf{j} = \mathbf{a} + 2\mathbf{b} + 2\mathbf{i}$$
$$-\mathbf{i} + \mathbf{j} = \mathbf{a} + 2\mathbf{b}$$

From equation (2): $\mathbf{a} = \mathbf{i} - \mathbf{j}$

so

$$-\mathbf{i} + \mathbf{j} = \mathbf{i} - \mathbf{j} + 2\mathbf{b}$$
$$2\mathbf{b} = -2\mathbf{i} + 2\mathbf{j}$$
$$\mathbf{b} = -\mathbf{i} + \mathbf{j}$$

Hence:

$$\mathbf{r} = \mathrm{e}^t[(\mathbf{i} - \mathbf{j})\cos 2t + (-\mathbf{i} + \mathbf{j})\sin 2t] + 2\mathbf{i}\sin t + \mathbf{i}\cos t$$

Comparing the above solution with the solution for example 2 shows that determining the values of the arbitrary constants in the complementary function must be delayed until the general solution (complementary function + particular integral) has been obtained.

As with ordinary differential equations, the form of the particular integral will depend on the form of $\mathbf{g}(t)$. The following table shows the functions to try:

$\mathbf{g}(t)$	Try:
$\mathbf{p}$ (constant vector)	$\mathbf{r} = \mathbf{a}$ (constant vector)
$\mathbf{p}t + \mathbf{q}$	$\mathbf{r} = \mathbf{a}t + \mathbf{b}$
$\mathbf{p}e^{mt}$	$\mathbf{r} = \mathbf{a}e^{mt}$
$\mathbf{p}\cos mt + \mathbf{q}\sin mt$	$\mathbf{r} = \mathbf{a}\cos mt + \mathbf{b}\sin mt$

The solution of the vector equation $\frac{d\mathbf{r}}{dt} + f(t)\mathbf{r} = \mathbf{a}e^{bt}$ where a and b are constants

The solution of an equation of this type requires the use of an integrating factor (see Book P4, chapter 5).

The integrating factor needed here is $e^{\int f(t)dt}$.

Multiplying the differential equation on both sides by the integrating factor gives

$$e^{\int f(t)dt}\frac{d\mathbf{r}}{dt} + e^{\int f(t)dt} \cdot f(t) \cdot \mathbf{r} = \mathbf{a}e^{\int f(t)dt} \cdot e^{bt}$$

and so
$$\frac{d}{dt}\left(e^{\int f(t)dt} \cdot \mathbf{r}\right) = \mathbf{a}e^{\left(\int f(t)dt + bt\right)}$$

Hence, integrating with respect to time gives

$$\mathbf{r} \cdot e^{\int f(t)dt} = \mathbf{a}\int e^{\left(\int f(t)dt + bt\right)}dt$$

and therefore a solution can be obtained for $\mathbf{r}$.

Example 6

At time t seconds the position vector of a particle P relative to a fixed origin is $\mathbf{r}$ metres and satisfies the differential equation

$$\frac{d\mathbf{r}}{dt} + 5\mathbf{r} = (3\mathbf{i} + 2\mathbf{j} + \mathbf{k})e^t$$

Given that P is at O when $t = 0$, find an expression for $\mathbf{r}$ in terms of t.

$$\frac{d\mathbf{r}}{dt} + 5\mathbf{r} = (3\mathbf{i} + 2\mathbf{j} + \mathbf{k})e^t$$

The integrating factor required is

$$e^{\int 5dt} = e^{5t}$$

Multiplying the equation by the integrating factor gives

$$e^{5t}\frac{d\mathbf{r}}{dt} + 5\mathbf{r}e^{5t} = (3\mathbf{i} + 2\mathbf{j} + \mathbf{k})e^{t} \cdot e^{5t}$$

$$\frac{d}{dt}(\mathbf{r}e^{5t}) = (3\mathbf{i} + 2\mathbf{j} + \mathbf{k})e^{6t}$$

Integrating with respect to t gives

$$\mathbf{r}e^{5t} = (3\mathbf{i} + 2\mathbf{j} + \mathbf{k}) \cdot \tfrac{1}{6}e^{6t} + \mathbf{c}$$

where $\mathbf{c}$ is a constant vector.

When $t = 0$, $\mathbf{r} = \mathbf{0}$ so

$$\mathbf{0} = \tfrac{1}{6}(3\mathbf{i} + 2\mathbf{j} + \mathbf{k}) + \mathbf{c}$$

$$\mathbf{c} = -\tfrac{1}{6}(3\mathbf{i} + 2\mathbf{j} + \mathbf{k})$$

and so $$\mathbf{r}e^{5t} = \tfrac{1}{6}(3\mathbf{i} + 2\mathbf{j} + \mathbf{k})e^{6t} - \tfrac{1}{6}(3\mathbf{i} + 2\mathbf{j} + \mathbf{k})$$

Dividing by e^{5t} to obtain an expression for $\mathbf{r}$ gives

$$\mathbf{r} = \tfrac{1}{6}(3\mathbf{i} + 2\mathbf{j} + \mathbf{k})(e^{t} - e^{-5t})$$

Exercise 1A

1 At time t seconds, the position vector of a particle P, relative to a fixed origin O, is $\mathbf{r}$ metres and its velocity is $\mathbf{v}\,\text{m s}^{-1}$. The particle moves so that $\mathbf{v}$ satisfies the differential equation $\frac{d\mathbf{v}}{dt} = 4\mathbf{v}$. Given that $\mathbf{r} = 4\mathbf{i}$ and $\mathbf{v} = 2\mathbf{i} - 3\mathbf{j}$ when $t = 0$, find $\mathbf{r}$ in terms of t.

2 A particle moves so that its velocity $\mathbf{v}\,\text{m s}^{-1}$ at time t seconds satisfies the differential equation $\frac{d\mathbf{v}}{dt} + 4\mathbf{v} = \mathbf{0}$. The position vector of the particle at time t seconds is $\mathbf{r}$ metres relative to a fixed origin O. Given that $\mathbf{r} = 3\mathbf{i} - 3\mathbf{j}$ and $\mathbf{v} = 2\mathbf{i}$ when $t = 0$, find

(a) $\mathbf{r}$ in terms of t,

(b) the distance of the particle from O when $t = 2$.

3 A particle moves so that its velocity $\mathbf{v}\,\text{m s}^{-1}$ at time t seconds satisfies the differential equation $\frac{d\mathbf{v}}{dt} = 2\mathbf{v}$. Given that $\mathbf{v} = 4\mathbf{i} + 2\mathbf{j}$ when $t = 0$ find

(a) $\mathbf{v}$ in terms of t,

(b) the speed of the particle when $t = 4$.

4 The position vector $\mathbf{r}$ metres of a particle, relative to a fixed origin O, at time t seconds satisfies the differential equation

$$\frac{d^2\mathbf{r}}{dt^2} + 2\frac{d\mathbf{r}}{dt} + 5\mathbf{r} = \mathbf{0}$$

Given that when $t = 0$, $\mathbf{r} = \mathbf{i} + \mathbf{j}$ and $\frac{d\mathbf{r}}{dt} = \mathbf{i}$, find $\mathbf{r}$ in terms of t and the distance of the particle from O when $t = \frac{\pi}{4}$.

5 The position vector $\mathbf{r}$ metres, relative to a fixed origin O, of a particle at time t seconds satisfies the differential equation

$$\frac{d^2\mathbf{r}}{dt^2} + 16\mathbf{r} = \mathbf{0}$$

Given that $\mathbf{r} = 2\mathbf{i} + 2\mathbf{j}$ and $\frac{d\mathbf{r}}{dt} = 4\mathbf{i} - 8\mathbf{j}$ when $t = 0$, find $\mathbf{r}$ in terms of t.

6 The position vector $\mathbf{r}$ metres, relative to a fixed origin O, of a particle at time t seconds satisfies the differential equation

$$\frac{d^2\mathbf{r}}{dt^2} + 2\frac{d\mathbf{r}}{dt} + \mathbf{r} = \mathbf{0}$$

Given that when $t = 0$, $\mathbf{r} = \mathbf{i} + 2\mathbf{j}$ and $\frac{d\mathbf{r}}{dt} = \mathbf{i} - \mathbf{j}$, find $\mathbf{r}$ in terms of t.

7 The position vector $\mathbf{r}$ metres, relative to a fixed origin O, of a particle P at time t seconds satisfies the differential equation

$$\frac{d\mathbf{r}}{dt} + 2\mathbf{r} = (15\mathbf{i} + 10\mathbf{j})e^{3t}$$

When $t = 0$, P is at the point with position vector $2\mathbf{i} + \mathbf{j}$. Find $\mathbf{r}$ in terms of t.

8 At time t seconds, the position vector of a particle P, relative to a fixed origin O, is $\mathbf{r}$ metres and its velocity is $\mathbf{v}\,\mathrm{m\,s^{-1}}$. The equation of motion for P is $\frac{d\mathbf{v}}{dt} = 2\mathbf{v}$. Given that $\mathbf{r} = 2\mathbf{i} + 4\mathbf{k}$ and $\mathbf{v} = \mathbf{i} + \mathbf{j} + \mathbf{k}$ when $t = 0$, find $\mathbf{r}$ in terms of t. Calculate the speed of P when $t = 2$.

9 The position vector **r** metres, relative to a fixed origin O, of a particle P at time t seconds satisfies the differential equation

$$\frac{d^2\mathbf{r}}{dt^2} + 4\frac{d\mathbf{r}}{dt} + 5\mathbf{r} = \mathbf{0}$$

Given that when $t = 0$, $\mathbf{r} = \mathbf{i} + 2\mathbf{j} + \mathbf{k}$ and $\frac{d\mathbf{r}}{dt} = \mathbf{k}$, find

(a) **r** in terms of t,

(b) the distance of the particle from O when $t = \frac{\pi}{2}$.

10 The position vector **r** metres, relative to a fixed origin O, of a particle P at time t seconds satisfies the differential equation

$$\frac{d\mathbf{r}}{dt} + \mathbf{r} = \mathbf{k}e^t$$

When $t = 0$, $r = 2\mathbf{i} + 3\mathbf{j} + \mathbf{k}$. Find **r** in terms of t.

11 The position vector **r** metres of a particle P, relative to a fixed origin O, at time t seconds, satisfies the vector differential equation

$$\frac{d^2\mathbf{r}}{dt^2} + 2\frac{d\mathbf{r}}{dt} + 5\mathbf{r} = \mathbf{0}$$

When $t = 0$, $\mathbf{r} = \mathbf{i} + \mathbf{j}$ and $\frac{d\mathbf{r}}{dt} = \mathbf{i} - \mathbf{j}$.

Find **r** in terms of t and hence find the smallest possible value of t for which P is moving parallel to **j**.

12 At time t seconds the position vector of a particle P, relative to a fixed origin O, is **r** metres, where **r** satisfies the differential equation

$$\frac{d^2\mathbf{r}}{dt^2} + 3\frac{d\mathbf{r}}{dt} = 0$$

At $t = 0$, P is at the point with position vector $(\mathbf{i} + 2\mathbf{k})$ m and has velocity $(3\mathbf{i} + 2\mathbf{j} + \mathbf{k})\ \text{m s}^{-1}$. Find an expression for **r** in terms of t.

13 At time t seconds the position vector of a particle P, relative to a fixed origin O, is **r** metres, where **r** satisfies the differential equation

$$\frac{d^2\mathbf{r}}{dt^2} - 4\frac{d\mathbf{r}}{dt} + 4\mathbf{r} = 8\mathbf{i}$$

At $t = 0$, P is at the point with position vector $(2\mathbf{i} - \mathbf{k})$ m and has velocity $(\mathbf{i} + 2\mathbf{j})\ \text{m s}^{-1}$. Find an expression for **r** in terms of t.

14 The position vector of a particle P, at time t seconds, relative to a fixed origin O, is $\mathbf{r}$ metres, where $\mathbf{r}$ satisfies the differential equation

$$\frac{\mathrm{d}^2\mathbf{r}}{\mathrm{d}t^2} - 4\mathbf{r} = 12\mathbf{i}t - 2\mathbf{j}$$

At $t = 0$, P is at the point with position vector $(\mathbf{i} + \mathbf{k})$ m and has velocity $2\mathbf{j}\,\mathrm{m\,s^{-1}}$. Find an expression for $\mathbf{r}$ in terms of t.

15 The position vector of a particle P, at time t seconds, relative to a fixed origin O, is $\mathbf{r}$ metres, where $\mathbf{r}$ satisfies the differential equation

$$\frac{\mathrm{d}\mathbf{r}}{\mathrm{d}t} + 3\mathbf{r} = 4\mathrm{e}^{-t}\mathbf{j}$$

When $t = 0$, $\mathbf{r} = 2\mathbf{i} - \mathbf{j}$. Find an expression for $\mathbf{r}$ in terms of t.

16 At time t seconds the position vector of a particle P, relative to a fixed origin O, is $\mathbf{r}$ metres, where $\mathbf{r}$ satisfies the differential equation

$$\frac{\mathrm{d}^2\mathbf{r}}{\mathrm{d}t^2} - 2\frac{\mathrm{d}\mathbf{r}}{\mathrm{d}t} - 8\mathbf{r} = (9\mathbf{i} + 18\mathbf{j})\mathrm{e}^{t}$$

At $t = 0$, P is at the point with position vector $(\mathbf{i} + 2\mathbf{j})$ m and has velocity $(2\mathbf{i} + \mathbf{j})\,\mathrm{m\,s^{-1}}$. Find an expression for $\mathbf{r}$ in terms of t.

1.2 Work done by a constant force

Consider a particle P which is being moved along a horizontal plane by a horizontal force $\mathbf{F}$ which is acting at an angle θ to the direction of motion of the particle.

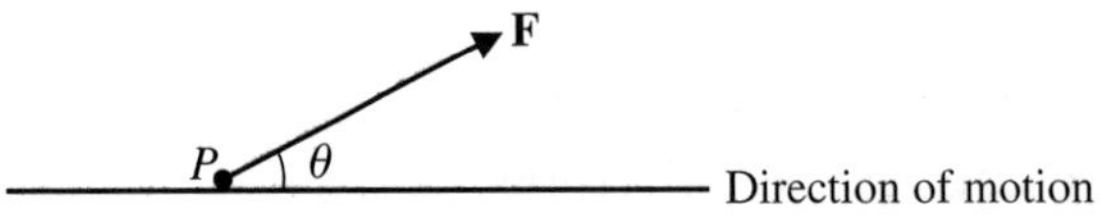

The work done by the force $\mathbf{F}$ as P moves a distance d along the plane is the scalar quantity given by $F\cos\theta \times d$ where F is the magnitude of $\mathbf{F}$.

Let the vector $\mathbf{d}$ be the displacement of P during this time. That is, d is the magnitude of $\mathbf{d}$.

Then: work done $= F\cos\theta \times d$ is equivalent to

■ **work done = F.d** (Book P3, chapter 5)

Example 7

A particle P is acted upon by a total force $\mathbf{F}$ where $\mathbf{F} = (5\mathbf{i} - 2\mathbf{j} + 3\mathbf{k})$ N. Calculate the work done by $\mathbf{F}$ as P moves from the point A with position vector $\mathbf{r}_A = (6\mathbf{i} + 5\mathbf{j} + \mathbf{k})$ metres relative to the origin, O, to the point B with position vector $\mathbf{r}_B = (8\mathbf{i} - 7\mathbf{j} + 6\mathbf{k})$ metres relative to O.

The displacement vector of P is $\overrightarrow{AB}$.

$$\begin{aligned}
\overrightarrow{AB} = \mathbf{r}_B - \mathbf{r}_A &= (8\mathbf{i} - 7\mathbf{j} + 6\mathbf{k}) - (6\mathbf{i} + 5\mathbf{j} + \mathbf{k}) \\
&= 2\mathbf{i} - 12\mathbf{j} + 5\mathbf{k} \\
\text{Work done} &= \mathbf{F}.\mathbf{d} \\
&= (5\mathbf{i} - 2\mathbf{j} + 3\mathbf{k}).(2\mathbf{i} - 12\mathbf{j} + 5\mathbf{k}) \\
&= 10 + 24 + 15 \\
&= 49
\end{aligned}$$

The work done by $\mathbf{F}$ as P moves from A to B is 49 J.

Example 8

A small smooth ring of mass 0.5 kg moves along a smooth, straight wire. The only forces acting on the ring are a constant force $\mathbf{F} = (5\mathbf{i} + 2\mathbf{j} - 3\mathbf{k})$ N and the normal contact force due to the wire. The ring is initially at rest at the point P with position vector $(\mathbf{i} + \mathbf{j} + \mathbf{k})$ m. Find the speed of the ring, in m s^{-1}, as it passes through the point Q with position vector $(3\mathbf{i} + 2\mathbf{j} - \mathbf{k})$ m.

The displacement vector of the ring is

$$\begin{aligned}
\mathbf{d} &= \mathbf{r}_Q - \mathbf{r}_P \\
&= (3\mathbf{i} + 2\mathbf{j} - \mathbf{k}) - (\mathbf{i} + \mathbf{j} + \mathbf{k}) \\
&= 2\mathbf{i} + \mathbf{j} - 2\mathbf{k} \\
\text{Work done} &= \mathbf{F}.\mathbf{d} \\
&= (5\mathbf{i} + 2\mathbf{j} - 3\mathbf{k}).(2\mathbf{i} + \mathbf{j} - 2\mathbf{k}) \\
&= 10 + 2 + 6 \\
&= 18
\end{aligned}$$

The work done by $\mathbf{F}$ is 18 J.

The normal contact force is at right angles to the direction of motion and therefore does no work.

By the work–energy principle, the gain in kinetic energy (K.E.) of the ring is therefore 18 J.

Initial K.E. of ring = 0

Final K.E. of ring $= \frac{1}{2}mv^2 = \frac{1}{2} \times 0.5v^2 = \frac{1}{4}v^2$

So
$$\frac{1}{4}v^2 = 18$$
$$v^2 = 72$$
$$v = 8.485$$

The speed of the ring as it passes through Q is $8.49\,\text{m}\,\text{s}^{-1}$.

1.3 The vector moment of a force

Consider a force **F** and let P be any point on the line of action of **F** with position vector **r** relative to the origin O. Let the angle between **r** and the line of action of **F** be θ.

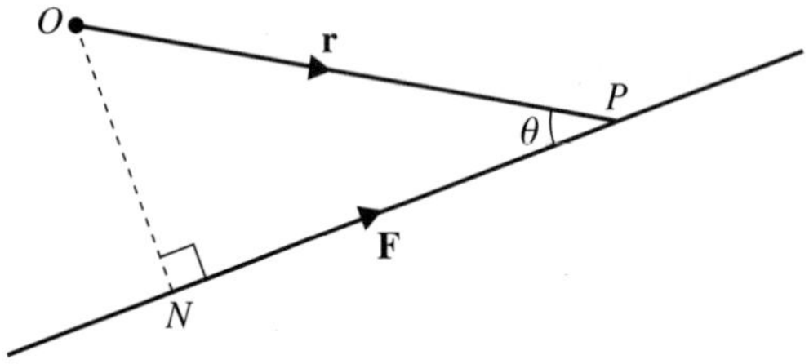

The moment of the force **F** about O has been defined in Book M1, chapter 6 to be

$$ON \times F = r \sin\theta \times F$$

where r and F are the magnitudes of **r** and **F** respectively.

But the magnitude of $\mathbf{r} \times \mathbf{F}$ is also $r \sin\theta \times F$. This suggests the following definition:

- **The vector moment of the force F about the origin O is given by:**

  vector moment of force = r × F

  where r is the position vector, relative to Q, of any point on the line of action of F.

Note that the moment of a force is a vector, that is, it has both magnitude and direction. It is directed along the perpendicular through O to the plane containing **r** and **F**. From the definition of the vector product, the direction of $\mathbf{r} \times \mathbf{F}$ along this perpendicular is in the sense of a right-handed screw, turned from **r** to **F**. This gives

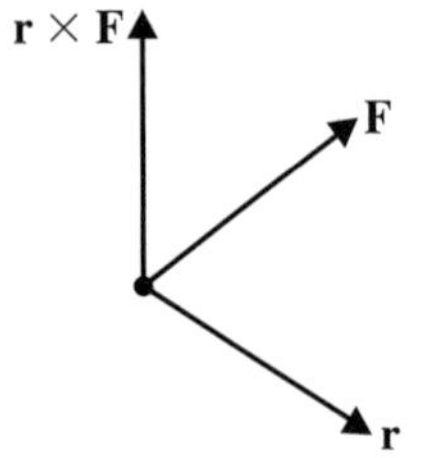

or

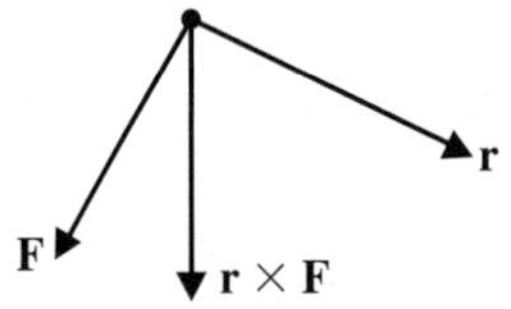

Example 9

The force $\mathbf{F}$, where $\mathbf{F} = (5\mathbf{i} - 2\mathbf{j} + 3\mathbf{k})$ N, acts through the point P whose position vector, relative to the origin O, is given by $\mathbf{r}_P = (2\mathbf{i} + 3\mathbf{j} + 4\mathbf{k})$ m.

Find the magnitude of the moment of $\mathbf{F}$

(a) about O,
(b) about the point A with position vector $\mathbf{r}_A = (3\mathbf{i} + 2\mathbf{j})$.

(a) Vector moment about $O = \mathbf{r}_P \times \mathbf{F}$

$$= (2\mathbf{i} + 3\mathbf{j} + 4\mathbf{k}) \times (5\mathbf{i} - 2\mathbf{j} + 3\mathbf{k})$$

$$= \begin{vmatrix} \mathbf{i} & \mathbf{j} & \mathbf{k} \\ 2 & 3 & 4 \\ 5 & -2 & 3 \end{vmatrix}$$

$$= 17\mathbf{i} + 14\mathbf{j} - 19\mathbf{k}$$

The magnitude of the vector moment $= |17\mathbf{i} + 14\mathbf{j} - 19\mathbf{k}|$

$$= \sqrt{(17^2 + 14^2 + 19^2)}$$

$$= 29.1$$

The magnitude of the vector moment of $\mathbf{F}$ about O is 29.1 N m.

(b) To find the vector moment of $\mathbf{F}$ about A, we first find the vector $\overrightarrow{AP}$.

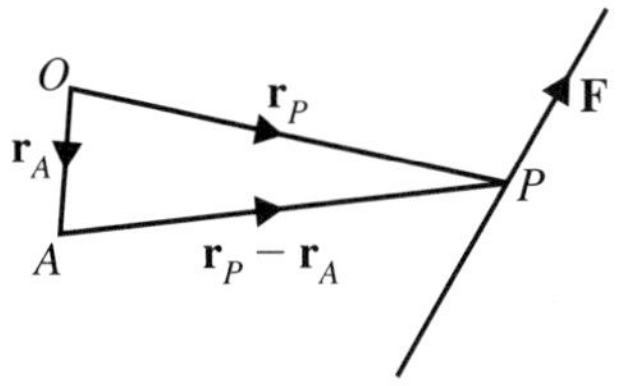

$$\overrightarrow{AP} = \mathbf{r}_P - \mathbf{r}_A$$

$$= (2\mathbf{i} + 3\mathbf{j} + 4\mathbf{k}) - (3\mathbf{i} + 2\mathbf{j})$$

$$= -\mathbf{i} + \mathbf{j} + 4\mathbf{k}$$

Vector moment of $\mathbf{F}$ about $A = \overrightarrow{AP} \times \mathbf{F}$

$$= (-\mathbf{i} + \mathbf{j} + 4\mathbf{k}) \times (5\mathbf{i} - 2\mathbf{j} + 3\mathbf{k})$$

$$= \begin{vmatrix} \mathbf{i} & \mathbf{j} & \mathbf{k} \\ -1 & 1 & 4 \\ 5 & -2 & 3 \end{vmatrix}$$

$$= 11\mathbf{i} + 23\mathbf{j} - 3\mathbf{k}$$

The magnitude of the vector moment $= \sqrt{(11^2 + 23^2 + 3^2)}$

$$= 25.7$$

The magnitude of the vector moment of $\mathbf{F}$ about A is 25.7 N m.

Exercise 1B

1 In each of the following problems, calculate the work done by the force $\mathbf{F}$ as it moves a particle through the displacement $\mathbf{r}$.

(a) $\mathbf{F} = (5\mathbf{i} + 7\mathbf{j} + 9\mathbf{k})\,\text{N}$, $\mathbf{r} = (-\mathbf{i} + 5\mathbf{j} - 2\mathbf{k})\,\text{m}$.

(b) $\mathbf{F} = (9\mathbf{i} - 2\mathbf{j} - 5\mathbf{k})\,\text{N}$, $\mathbf{r} = (2\mathbf{i} + 7\mathbf{j} - 3\mathbf{k})\,\text{m}$.

(c) $\mathbf{F} = (-2\mathbf{i} + 7\mathbf{j} - 3\mathbf{k})\,\text{N}$, $\mathbf{r} = (4\mathbf{i} + 2\mathbf{j} + \mathbf{k})\,\text{m}$.

2 In each of the following problems, calculate the work done by the force $\mathbf{F}$ as it moves a particle P from the point A with position vector $\mathbf{r}_A$ relative to the origin, O, to the point B with position vector $\mathbf{r}_B$ relative to O.

(a) $\mathbf{F} = (3\mathbf{i} + 2\mathbf{j} + \mathbf{k})\,\text{N}$, $\mathbf{r}_A = (6\mathbf{i} - 2\mathbf{j} + \mathbf{k})\,\text{m}$, $\mathbf{r}_B = (7\mathbf{i} + 6\mathbf{j} + 5\mathbf{k})\,\text{m}$.

(b) $\mathbf{F} = (4\mathbf{i} - \mathbf{j} - 2\mathbf{k})\,\text{N}$, $\mathbf{r}_A = (\mathbf{i} + 2\mathbf{j} + 3\mathbf{k})\,\text{m}$, $\mathbf{r}_B = (10\mathbf{i} + 8\mathbf{j} + 6\mathbf{k})\,\text{m}$.

(c) $\mathbf{F} = (-2\mathbf{i} + 7\mathbf{j} - \mathbf{k})\,\text{N}$, $\mathbf{r}_A = (3\mathbf{i} + 2\mathbf{j} + 3\mathbf{k})\,\text{m}$, $\mathbf{r}_B = (\mathbf{i} + 2\mathbf{j} - \mathbf{k})\,\text{m}$.

3 Given that the particle P in question 2 has mass 2.5 kg and that P starts from rest at point A, calculate in each case the speed of P as it passes through the point B. You may assume that the force $\mathbf{F}$ is the only force doing work on P during the motion.

4 A particle P is acted upon by three forces of magnitude 5 N, 4 N and 10 N acting in the directions of the vectors $2\mathbf{i} + \mathbf{j} + 2\mathbf{k}$, $6\mathbf{i} + 8\mathbf{j}$ and $\mathbf{i} + 8\mathbf{j} + 4\mathbf{k}$ respectively. The three forces cause a displacement of $(7\mathbf{i} - 2\mathbf{j})\,\text{m}$. Find

(a) the resultant of the three forces,

(b) the total work done by the three forces.

5 A particle P of mass 0.5 kg is initially at rest at the point with position vector $(3\mathbf{i} + 2\mathbf{j} + 4\mathbf{k})\,\text{m}$. A constant force is applied to P and 2 s later P passes through the point with position vector $(9\mathbf{i} + 12\mathbf{j} + 8\mathbf{k})$. Find

(a) the acceleration of P,

(b) the constant force,

(c) the work done by the force during this time interval.

6 A particle P is acted upon by a constant force of magnitude 2 N parallel to the vector $(2\mathbf{i} + 6\mathbf{j} + 3\mathbf{k})$ and a constant force of magnitude 3 N parallel to the vector $(2\mathbf{i} - 2\mathbf{j} + \mathbf{k})$. P is initially at the point with position vector $(2\mathbf{i} - 9\mathbf{k})$ and is moved to the point with position vector $(6\mathbf{i} + \mathbf{j} - 2\mathbf{k})$. Calculate the total work done by the two forces.

7 In each of the following problems find the vector moment about O of the force $\mathbf{F}$ which passes through the point with position vector $\mathbf{r}$ relative to O.

(a) $\mathbf{F} = (5\mathbf{i} + 6\mathbf{j} + 7\mathbf{k})$ N, $\quad \mathbf{r} = (2\mathbf{i} - 2\mathbf{j} + 2\mathbf{k})$ m.

(b) $\mathbf{F} = (-9\mathbf{i} + 2\mathbf{j} + 5\mathbf{k})$ N, $\quad \mathbf{r} = (3\mathbf{i} + 7\mathbf{j} - 2\mathbf{k})$ m.

(c) $\mathbf{F} = (2\mathbf{i} - 7\mathbf{j} - 4\mathbf{k})$ N, $\quad \mathbf{r} = (4\mathbf{i} - 2\mathbf{j} - 3\mathbf{k})$ m.

8 In each of the following problems, find the vector moment about the point P of the force $\mathbf{F}$ which passes through the point with position vector $\mathbf{r}$ relative to the origin O.

(a) $\mathbf{F} = (5\mathbf{i} + 6\mathbf{j} + 7\mathbf{k})$ N, $\quad \mathbf{r} = (2\mathbf{i} - 2\mathbf{j} + 2\mathbf{k})$ m, $\quad P$ is (2, 3, 4).

(b) $\mathbf{F} = (4\mathbf{i} - 2\mathbf{j} + 3\mathbf{k})$ N, $\quad \mathbf{r} = (3\mathbf{i} - 7\mathbf{j} + 3\mathbf{k})$ m, $\quad P$ is (1, 0, −1).

(c) $\mathbf{F} = (3\mathbf{i} - 3\mathbf{j} + 2\mathbf{k})$ N, $\quad \mathbf{r} = (4\mathbf{i} - 4\mathbf{j} - 2\mathbf{k})$ m, $\quad P$ is (1, −2, 3).

9 The line of action of a force $\mathbf{F}$ of magnitude 15 N is parallel to the vector $3\mathbf{i} + 4\mathbf{k}$ and passes through the point with position vector $(3\mathbf{i} + 3\mathbf{j} - 4\mathbf{k})$ m relative to the origin O. Find the vector moment of $\mathbf{F}$

(a) about the origin,

(b) about the point with position vector $(5\mathbf{i} + 3\mathbf{j} + \mathbf{k})$ m relative to O.

10 The force $\mathbf{F}$ is given by $\mathbf{F} = (2\mathbf{i} + 3\mathbf{j} + 2\mathbf{k})$ N. The vector moment of $\mathbf{F}$ about the origin is $(5\mathbf{i} - 14\mathbf{j} + 16\mathbf{k})$ N m. A point A on the line of action of $\mathbf{F}$ has position vector $(x\mathbf{i} + y\mathbf{j} + z\mathbf{k})$ m relative to O where x, y and z are integers. The distance OA is 9 m. Find the values of x, y and z.

1.4 The analysis of systems of forces

The simplest case of a system of forces acting on a rigid body arises when the forces all act through a common point. In such a case, the resultant of the forces is found by adding the forces.

Example 10

Forces $\mathbf{F}_1$, $\mathbf{F}_2$ and $\mathbf{F}_3$, where $\mathbf{F}_1 = (-3\mathbf{i} + 6\mathbf{j} + 7\mathbf{k})$ N, $\mathbf{F}_2 = (5\mathbf{i} + 7\mathbf{j} - 2\mathbf{k})$ N and $\mathbf{F}_3 = (9\mathbf{i} - 5\mathbf{j} + 3\mathbf{k})$ N, all act at point P of a rigid body.

(a) Find the resultant of the three forces.

A fourth force $\mathbf{F}_4$, also acting through P, is to be introduced and the resulting system is to be in equilibrium.

(b) Find the force $\mathbf{F}_4$.

(a) Resultant $= \mathbf{F}_1 + \mathbf{F}_2 + \mathbf{F}_3$

$$= (-3\mathbf{i} + 6\mathbf{j} + 7\mathbf{k}) + (5\mathbf{i} + 7\mathbf{j} - 2\mathbf{k}) + (9\mathbf{i} - 5\mathbf{j} + 3\mathbf{k})$$
$$= 11\mathbf{i} + 8\mathbf{j} + 8\mathbf{k}$$

The resultant is $(11\mathbf{i} + 8\mathbf{j} + 8\mathbf{k})$ N.

(b) For the new system to be in equilibrium,

$$\mathbf{F}_1 + \mathbf{F}_2 + \mathbf{F}_3 + \mathbf{F}_4 = \mathbf{0}$$

or

$$\mathbf{F}_4 = -(\mathbf{F}_1 + \mathbf{F}_2 + \mathbf{F}_3)$$

The new force $\mathbf{F}_4$ is $-(11\mathbf{i} + 8\mathbf{j} + 8\mathbf{k})$ N.

Couples

When the forces acting on a body do not all act through the same point, it is possible for the sum of the forces to be zero, so there is no resultant force, although the sum of the moments of the separate forces about some point is not zero. To illustrate this, consider the case of a pair of forces **F** and $-\mathbf{F}$ acting on a body and having different lines of action. Such a pair will act in opposite directions along a pair of parallel lines, as shown in the diagram:

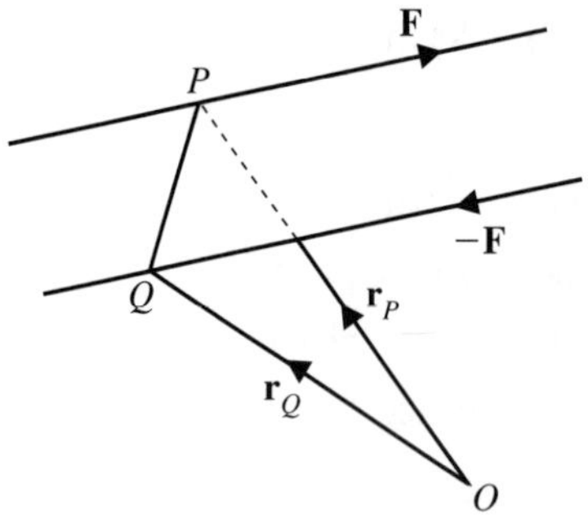

Let O be any fixed point which need not be in the plane of **F** and $-\mathbf{F}$. Let P and Q be any points on the lines of action of **F** and $-\mathbf{F}$ respectively, having position vectors $\mathbf{r}_P$ and $\mathbf{r}_Q$ relative to O. Then the vector moment of **F** about O is $\mathbf{r}_P \times \mathbf{F}$ and the vector moment of $-\mathbf{F}$ about O is $\mathbf{r}_Q \times (-\mathbf{F})$. The total vector moment about O of the system

$$= (\mathbf{r}_P \times \mathbf{F}) + [\mathbf{r}_Q \times (-\mathbf{F})]$$
$$= (\mathbf{r}_P \times \mathbf{F}) - (\mathbf{r}_Q \times \mathbf{F})$$
$$= (\mathbf{r}_P - \mathbf{r}_Q) \times \mathbf{F}$$
$$= \overrightarrow{QP} \times \mathbf{F}$$

Thus the vector moment of the pair of forces is independent of the position of O. Consequently, the vector moment of the pair of forces is the same about any point. Such a pair of forces is called a **couple**.

Equivalent systems of forces

Two systems of forces are said to be **equivalent** if they have the same effect when applied to the same body. A system of coplanar forces must be equivalent to

(a) a single resultant force, or
(b) to a couple, or
(c) must be in equilibrium.

When forces in three dimensions are being considered, another possibility arises.

Poinsot's reduction of a system of forces

Consider a set of forces $\mathbf{F}_1$, $\mathbf{F}_2$, $\mathbf{F}_3$, ..., $\mathbf{F}_n$ acting on a body with their lines of action passing through the points with position vectors $\mathbf{r}_1$, $\mathbf{r}_2$, $\mathbf{r}_3$, ..., $\mathbf{r}_n$ relative to a fixed origin O. Corresponding to each force $\mathbf{F}_1$, $\mathbf{F}_2$, $\mathbf{F}_3$, ..., $\mathbf{F}_n$, introduce at O pairs of forces $\pm\mathbf{F}_1$, $\pm\mathbf{F}_2$, $\pm\mathbf{F}_3$, ..., $\pm\mathbf{F}_n$.

For a typical force $\mathbf{F}_i$ we have

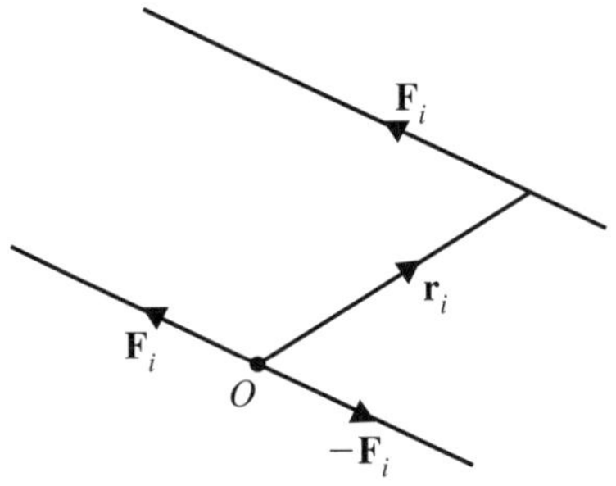

The resultant of the system of forces is unchanged by the addition of these pairs of forces. However, as can be seen from the diagram, the system now consists of a set of forces $\mathbf{F}_1$, $\mathbf{F}_2$, $\mathbf{F}_3$, ..., $\mathbf{F}_n$ acting at O together with a set of couples of moment

$$\mathbf{r}_1 \times \mathbf{F}_1, \mathbf{r}_2 \times \mathbf{F}_2, \mathbf{r}_3 \times \mathbf{F}_3, \ldots, \mathbf{r}_n \times \mathbf{F}_n$$

Let the resultant of the forces $\mathbf{F}_1$, $\mathbf{F}_2$, $\mathbf{F}_3$, ..., $\mathbf{F}_n$ acting at O be $\mathbf{F}$.

Then $$\mathbf{F} = \mathbf{F}_1 + \mathbf{F}_2 + \mathbf{F}_3 + \ldots + \mathbf{F}_n$$

or $$\mathbf{F} = \sum_{i=1}^{n} \mathbf{F}_i$$

and the couples are equivalent to a single couple $\mathbf{G}$ where

$$\mathbf{G} = (\mathbf{r}_1 \times \mathbf{F}_1) + (\mathbf{r}_2 \times \mathbf{F}_2) + (\mathbf{r}_3 \times \mathbf{F}_3) + \ldots + (\mathbf{r}_n \times \mathbf{F}_n)$$

or $$\mathbf{G} = \sum_{i=1}^{n} \mathbf{r}_i \times \mathbf{F}_i$$

Thus:

- **A system of forces $\mathbf{F}_1$, $\mathbf{F}_2$, ..., $\mathbf{F}_n$ with lines of action passing through the points with position vectors $\mathbf{r}_1, \mathbf{r}_2, \ldots, \mathbf{r}_n$ relative to an origin O is equivalent to a single force $\mathbf{F} = \sum_{i=1}^{n} \mathbf{F}_i$ acting at O and a couple of moment $\mathbf{G} = \sum_{i=1}^{n} \mathbf{r}_i \times \mathbf{F}_i$.**

The resultant force $\mathbf{F}$ is the same, whatever point is taken to be the origin O. However, the moment of the couple changes if $\mathbf{F}$ is required to act at a point other than O. This can be seen by considering a system which reduces to a single force $\mathbf{F} = \sum_{i=1}^{n} \mathbf{F}_i$ at O and a couple of moment $\mathbf{G} = \sum_{i=1}^{n} \mathbf{r}_i \times \mathbf{F}_i$ about O.

Suppose instead that the resultant force is required to act through the point A with position vector $\mathbf{r}_A$ relative to O.

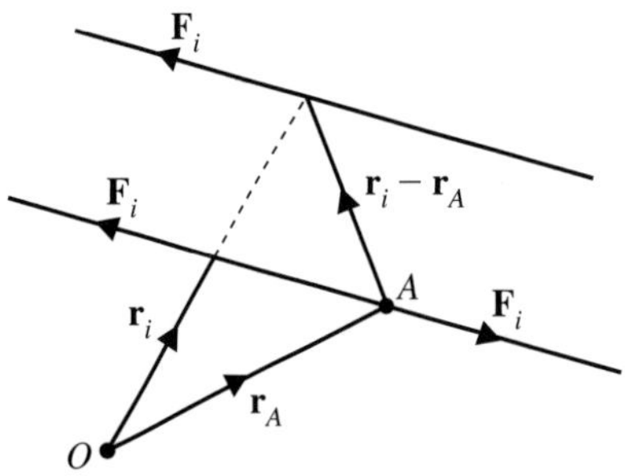

Pairs of forces $\pm\mathbf{F}_1$, $\pm\mathbf{F}_2$, ..., $\pm\mathbf{F}_n$ are introduced at A. This gives a set of forces $\mathbf{F}_1$, $\mathbf{F}_2$, ..., $\mathbf{F}_n$ acting at A, together with a set of couples of moment $(\mathbf{r}_1 - \mathbf{r}_A) \times \mathbf{F}_1$, $(\mathbf{r}_2 - \mathbf{r}_A) \times \mathbf{F}_2$, $\ldots (\mathbf{r}_n - \mathbf{r}_A) \times \mathbf{F}_n$.

Hence the resultant force $\mathbf{F}$ is given by

$$\mathbf{F} = \mathbf{F}_1 + \mathbf{F}_2 + \mathbf{F}_3 + \ldots + \mathbf{F}_n$$

which is the same as before but now it is acting at A. However, the couple is now

$$\begin{aligned}\sum_{i=1}^{n} (\mathbf{r}_i - \mathbf{r}_A) \times \mathbf{F}_i &= \left(\sum_{i=1}^{n} \mathbf{r}_i \times \mathbf{F}_i\right) - \left(\sum_{i=1}^{n} \mathbf{r}_A \times \mathbf{F}_i\right) \\ &= \mathbf{G} - \left(\mathbf{r}_A \times \sum_{i=1}^{n} \mathbf{F}_i\right) \\ &= \mathbf{G} - (\mathbf{r}_A \times \mathbf{F})\end{aligned}$$

All systems of forces can be reduced to a single force $\mathbf{F}$ and a couple of moment $\mathbf{G}$. Different cases arise when either or both of $\mathbf{F}$ and $\mathbf{G}$ are the zero vector.

(a) When $\mathbf{F} = \mathbf{0}$ the system reduces to a couple.
(b) When $\mathbf{G} = \mathbf{0}$ the system reduces to a single force.
(c) When $\mathbf{F} = \mathbf{0}$ and $\mathbf{G} = \mathbf{0}$, the system is in equilibrium.

As the moment of the couple about the point A, with position vector $\mathbf{r}_A$ relative to O, is $\mathbf{G} - (\mathbf{r}_A \times \mathbf{F})$ (see above) it follows that when both $\mathbf{F}$ and $\mathbf{G}$ are zero, the moment of the couple about the point A is also zero. Hence to prove a system is in equilibrium it is sufficient to show that

(a) $\mathbf{F} = \mathbf{0}$,
(b) the sum of the moments of the forces about *any* point is zero.

Alternatively, you can show that $\mathbf{F} = \mathbf{0}$ and the forces are concurrent, which implies a zero total moment about the common point.

Hence:

- **A system of forces in three dimensions is either in equilibrium or can be reduced to a single force, a couple or a couple and a force.**

Example 11

A rigid body is in equilibrium under the action of four forces $\mathbf{F}_1$, $\mathbf{F}_2$, $\mathbf{F}_3$ and $\mathbf{F}_4$ acting at points with position vectors $\mathbf{r}_1$, $\mathbf{r}_2$, $\mathbf{r}_3$ and $\mathbf{r}_4$ relative to a fixed origin O, where

$$\mathbf{F}_1 = (5\mathbf{i} + 2\mathbf{j} - \mathbf{k})\,\text{N} \qquad \mathbf{r}_1 = 4\mathbf{k}\,\text{m}$$

$$\mathbf{F}_2 = (3\mathbf{i} - 2\mathbf{j} - 2\mathbf{k})\,\text{N} \qquad \mathbf{r}_2 = (8\mathbf{i} + \mathbf{k})\,\text{m}$$

$$\mathbf{F}_3 = (-2\mathbf{i} + 3\mathbf{j} + 2\mathbf{k})\,\text{N} \qquad \mathbf{r}_3 = (\mathbf{i} + 8\mathbf{j} + 7\mathbf{k})\,\text{m}$$

Find $\mathbf{F}_4$ and an equation for the line of action of $\mathbf{F}_4$.

As the system is in equilibrium $\sum_i \mathbf{F}_i = \mathbf{0}$ and $\sum_i \mathbf{r}_i \times \mathbf{F}_i = \mathbf{0}$

So $\quad \mathbf{F}_1 + \mathbf{F}_2 + \mathbf{F}_3 + \mathbf{F}_4 = \mathbf{0}$

$$\begin{aligned}\mathbf{F}_4 &= -(\mathbf{F}_1 + \mathbf{F}_2 + \mathbf{F}_3)\\ &= -[(5\mathbf{i} + 2\mathbf{j} - \mathbf{k}) + (3\mathbf{i} - 2\mathbf{j} - 2\mathbf{k}) + (-2\mathbf{i} + 3\mathbf{j} + 2\mathbf{k})]\\ &= -6\mathbf{i} - 3\mathbf{j} + \mathbf{k}\end{aligned}$$

So $\mathbf{F}_4$ is $(-6\mathbf{i} - 3\mathbf{j} + \mathbf{k})\,\text{N}$.

The line of action of $\mathbf{F}_4$ is parallel to the vector $-6\mathbf{i} - 3\mathbf{j} + \mathbf{k}$, but to find its equation, the position vector $\mathbf{r}_4$ of a point on the line is required.

Since $$\sum_i \mathbf{r}_i \times \mathbf{F}_i = \mathbf{0}$$

it follows that $\mathbf{r}_4 \times \mathbf{F}_4 = -[(\mathbf{r}_1 \times \mathbf{F}_1) + (\mathbf{r}_2 \times \mathbf{F}_2) + (\mathbf{r}_3 \times \mathbf{F}_3)]$.

$$\mathbf{r}_1 \times \mathbf{F}_1 = \begin{vmatrix} \mathbf{i} & \mathbf{j} & \mathbf{k} \\ 0 & 0 & 4 \\ 5 & 2 & -1 \end{vmatrix} = -8\mathbf{i} + 20\mathbf{j}$$

$$\mathbf{r}_2 \times \mathbf{F}_2 = \begin{vmatrix} \mathbf{i} & \mathbf{j} & \mathbf{k} \\ 8 & 0 & 1 \\ 3 & -2 & -2 \end{vmatrix} = 2\mathbf{i} + 19\mathbf{j} - 16\mathbf{k}$$

$$\mathbf{r}_3 \times \mathbf{F}_3 = \begin{vmatrix} \mathbf{i} & \mathbf{j} & \mathbf{k} \\ 1 & 8 & 7 \\ -2 & 3 & 2 \end{vmatrix} = -5\mathbf{i} - 16\mathbf{j} + 19\mathbf{k}$$

So $\quad (\mathbf{r}_1 \times \mathbf{F}_1) + (\mathbf{r}_2 \times \mathbf{F}_2) + (\mathbf{r}_3 \times \mathbf{F}_3) = -11\mathbf{i} + 23\mathbf{j} + 3\mathbf{k}$

and hence
$$\mathbf{r}_4 \times \mathbf{F}_4 = -(-11\mathbf{i} + 23\mathbf{j} + 3\mathbf{k})$$
$$= 11\mathbf{i} - 23\mathbf{j} - 3\mathbf{k} \qquad (1)$$

Let $\mathbf{r}_4 = x\mathbf{i} + y\mathbf{j} + z\mathbf{k}$.

Then

$$\mathbf{r}_4 \times \mathbf{F}_4 = \begin{vmatrix} \mathbf{i} & \mathbf{j} & \mathbf{k} \\ x & y & z \\ -6 & -3 & 1 \end{vmatrix}$$
$$= (y + 3z)\mathbf{i} - (x + 6z)\mathbf{j} + (-3x + 6y)\mathbf{k} \qquad (2)$$

Equating the coefficients of **i**, **j** and **k** in equations (1) and (2) gives

$$(y + 3z) = 11 \qquad (3)$$
$$-(x + 6z) = -23 \qquad (4)$$
$$(-3x + 6y) = -3 \qquad (5)$$

Note that these equations do not have a unique solution as (x, y, z) are the coordinates of *any* point on the line of action of $\mathbf{F}_4$.

In particular, take $x = 0$.

Then from equation (4), $z = \frac{23}{6}$.

From equation (5) $y = -\frac{1}{2}$.

(These values of y and z also satisfy equation (3). If they did not, the system would not be in equilibrium or the line of action of $\mathbf{F}_4$ would not intersect the plane $x = 0$.)

Since the line of action of $\mathbf{F}_4$ is parallel to the vector $-6\mathbf{i} - 3\mathbf{j} + \mathbf{k}$ and passes through the point with position vector $-\frac{1}{2}\mathbf{j} + \frac{23}{6}\mathbf{k}$, it has equation

$$\mathbf{r} = -\tfrac{1}{2}\mathbf{j} + \tfrac{23}{6}\mathbf{k} + \lambda(-6\mathbf{i} - 3\mathbf{j} + \mathbf{k})$$

Example 12

Forces $\mathbf{F}_1$, $\mathbf{F}_2$ and $\mathbf{F}_3$ act on a body at points with position vectors $\mathbf{r}_1$, $\mathbf{r}_2$, $\mathbf{r}_3$ relative to a fixed origin O, where

$$\mathbf{F}_1 = (2\mathbf{i} + 3\mathbf{j} + 4\mathbf{k})\,\text{N} \qquad \mathbf{r}_1 = (-\mathbf{i} + 2\mathbf{j} + \mathbf{k})\,\text{m}$$
$$\mathbf{F}_2 = (4\mathbf{i} - 3\mathbf{j} + 2\mathbf{k})\,\text{N} \qquad \mathbf{r}_2 = (3\mathbf{i} - 3\mathbf{j})\,\text{m}$$
$$\mathbf{F}_3 = (-3\mathbf{i} + 2\mathbf{j} - \mathbf{k})\,\text{N} \qquad \mathbf{r}_3 = (4\mathbf{i} + 2\mathbf{k})\,\text{m}$$

This system can be reduced to a single force $\mathbf{F}$ acting at point A with position vector $\mathbf{r}_A = (\mathbf{i} + \mathbf{j} + \mathbf{k})$ together with a couple $\mathbf{G}$.

(a) Find the force $\mathbf{F}$ and the couple $\mathbf{G}$.
(b) Give, in vector form, an equation for the line of action of $\mathbf{F}$.

(a)
$$\mathbf{F} = \sum_i \mathbf{F}_i$$
$$= (2\mathbf{i} + 3\mathbf{j} + 4\mathbf{k}) + (4\mathbf{i} - 3\mathbf{j} + 2\mathbf{k}) + (-3\mathbf{i} + 2\mathbf{j} - \mathbf{k})$$
$$= 3\mathbf{i} + 2\mathbf{j} + 5\mathbf{k}$$

Take moments about A:

moment of $\mathbf{F}_i = (\mathbf{r}_i - \mathbf{r}_A) \times \mathbf{F}_i$

$$\text{moment of } \mathbf{F}_1 = \begin{vmatrix} \mathbf{i} & \mathbf{j} & \mathbf{k} \\ -2 & 1 & 0 \\ 2 & 3 & 4 \end{vmatrix} = 4\mathbf{i} + 8\mathbf{j} - 8\mathbf{k}$$

$$\text{moment of } \mathbf{F}_2 = \begin{vmatrix} \mathbf{i} & \mathbf{j} & \mathbf{k} \\ 2 & -4 & -1 \\ 4 & -3 & 2 \end{vmatrix} = -11\mathbf{i} - 8\mathbf{j} + 10\mathbf{k}$$

$$\text{moment of } \mathbf{F}_3 = \begin{vmatrix} \mathbf{i} & \mathbf{j} & \mathbf{k} \\ 3 & -1 & 1 \\ -3 & 2 & -1 \end{vmatrix} = -\mathbf{i} + 3\mathbf{k}$$

So
$$\mathbf{G} = \sum_i (\mathbf{r}_i - \mathbf{r}_A) \times \mathbf{F}_i$$
$$= (4\mathbf{i} + 8\mathbf{j} - 8\mathbf{k}) + (-11\mathbf{i} - 8\mathbf{j} + 10\mathbf{k}) + (-\mathbf{i} + 3\mathbf{k})$$
$$= -8\mathbf{i} + 5\mathbf{k}$$

The force is $(3\mathbf{i} + 2\mathbf{j} + 5\mathbf{k})\,\text{N}$ and the couple is $(-8\mathbf{i} + 5\mathbf{k})\,\text{N m}$.

(b) The line of action of $\mathbf{F}$ is parallel to $(3\mathbf{i} + 2\mathbf{j} + 5\mathbf{k})$ and passes through A where $\mathbf{r}_A = \mathbf{i} + \mathbf{j} + \mathbf{k}$.

An equation for the line of action of $\mathbf{F}$ is therefore

$$\mathbf{r} = \mathbf{i} + \mathbf{j} + \mathbf{k} + \lambda(3\mathbf{i} + 2\mathbf{j} + 5\mathbf{k})$$

Example 13

Three forces $\mathbf{F}_1$, $\mathbf{F}_2$ and $\mathbf{F}_3$ act on a rigid body at points having position vectors $\mathbf{r}_1$, $\mathbf{r}_2$ and $\mathbf{r}_3$ relative to a fixed origin O where

$$\mathbf{F}_1 = (-3\mathbf{i} - 2\mathbf{j} + 4\mathbf{k})\,\text{N} \qquad \mathbf{r}_1 = (2\mathbf{i} + 4\mathbf{j} - \mathbf{k})\,\text{m}$$
$$\mathbf{F}_2 = (5\mathbf{i} + 3\mathbf{j} - 7\mathbf{k})\,\text{N} \qquad \mathbf{r}_2 = (\mathbf{i} - \mathbf{j} - \mathbf{k})\,\text{m}$$
$$\mathbf{F}_3 = (-2\mathbf{i} - \mathbf{j} + 3\mathbf{k})\,\text{N} \qquad \mathbf{r}_3 = (2\mathbf{i} + 3\mathbf{k})\,\text{m}$$

Show that the system reduces to a couple, and find the magnitude of the couple.

If the system reduces to a couple, $\sum_i \mathbf{F}_i = \mathbf{0}$.

$$\sum_i \mathbf{F}_i = (-3\mathbf{i} - 2\mathbf{j} + 4\mathbf{k}) + (5\mathbf{i} + 3\mathbf{j} - 7\mathbf{k}) + (-2\mathbf{i} - \mathbf{j} + 3\mathbf{k}) = \mathbf{0}$$

So the system is either in equilibrium or reduces to a couple. To ascertain which of these is the case, the moments of the forces about O must be calculated.

$$\text{moment of } \mathbf{F}_1 = \mathbf{r}_1 \times \mathbf{F}_1 = \begin{vmatrix} \mathbf{i} & \mathbf{j} & \mathbf{k} \\ 2 & 4 & -1 \\ -3 & -2 & 4 \end{vmatrix} = 14\mathbf{i} - 5\mathbf{j} + 8\mathbf{k}$$

$$\text{moment of } \mathbf{F}_2 = \mathbf{r}_2 \times \mathbf{F}_2 = \begin{vmatrix} \mathbf{i} & \mathbf{j} & \mathbf{k} \\ 1 & -1 & -1 \\ 5 & 3 & -7 \end{vmatrix} = 10\mathbf{i} + 2\mathbf{j} + 8\mathbf{k}$$

$$\text{moment of } \mathbf{F}_3 = \mathbf{r}_3 \times \mathbf{F}_3 = \begin{vmatrix} \mathbf{i} & \mathbf{j} & \mathbf{k} \\ 2 & 0 & 3 \\ -2 & -1 & 3 \end{vmatrix} = 3\mathbf{i} - 12\mathbf{j} - 2\mathbf{k}$$

So

$$\sum_i \mathbf{r}_i \times \mathbf{F}_i = (14\mathbf{i} - 5\mathbf{j} + 8\mathbf{k}) + (10\mathbf{i} + 2\mathbf{j} + 8\mathbf{k}) + (3\mathbf{i} - 12\mathbf{j} - 2\mathbf{k})$$
$$= 27\mathbf{i} - 15\mathbf{j} + 14\mathbf{k}$$

As $\sum_i \mathbf{r}_i \times \mathbf{F}_i \neq \mathbf{0}$, the system reduces to a couple.

$$\text{The magnitude of the couple} = \sqrt{(27^2 + 15^2 + 14^2)}$$
$$= 33.9$$

The magnitude of the couple is 33.9 Nm.

Exercise 1C

1 Forces $\mathbf{F}_1$, $\mathbf{F}_2$, $\mathbf{F}_3$ and $\mathbf{F}_4$ act on a particle. Find the resultant of the forces in each of the following cases:

(a) $\mathbf{F}_1 = (5\mathbf{i} + 2\mathbf{j} - 3\mathbf{k})\,\text{N}$, $\mathbf{F}_2 = (3\mathbf{i} - 7\mathbf{k})\,\text{N}$,
$\mathbf{F}_3 = (-6\mathbf{i} + 2\mathbf{j} + 3\mathbf{k})\,\text{N}$, $\mathbf{F}_4 = (4\mathbf{i} + 2\mathbf{k})\,\text{N}$.

(b) $\mathbf{F}_1 = (9\mathbf{i} + 7\mathbf{j} - 3\mathbf{k})\,\text{N}$, $\mathbf{F}_2 = (4\mathbf{i} - 3\mathbf{j} + 5\mathbf{k})\,\text{N}$,
$\mathbf{F}_3 = (-3\mathbf{i} + 2\mathbf{j} - \mathbf{k})\,\text{N}$, $\mathbf{F}_4 = (3\mathbf{i} - 2\mathbf{j} + \mathbf{k})\,\text{N}$.

2 A particle is in equilibrium under the action of four forces $\mathbf{F}_1$, $\mathbf{F}_2$, $\mathbf{F}_3$ and $\mathbf{F}_4$. Find $\mathbf{F}_4$ when $\mathbf{F}_1$, $\mathbf{F}_2$ and $\mathbf{F}_3$ are as given:

(a) $\mathbf{F}_1 = (2\mathbf{i} - 3\mathbf{j})\,\text{N}$, $\mathbf{F}_2 = (5\mathbf{i} + 7\mathbf{j})\,\text{N}$, $\mathbf{F}_3 = (-3\mathbf{i} + \mathbf{j})\,\text{N}$.

(b) $\mathbf{F}_1 = (6\mathbf{i} + 2\mathbf{j} + \mathbf{k})\,\text{N}$, $\mathbf{F}_2 = (4\mathbf{i} + 3\mathbf{j} - 6\mathbf{k})\,\text{N}$, $\mathbf{F}_3 = (-\mathbf{i} - 2\mathbf{j} + 2\mathbf{k})\,\text{N}$.

(c) $\mathbf{F}_1 = (-9\mathbf{i} + 7\mathbf{j} + 5\mathbf{k})\,\text{N}$, $\mathbf{F}_2 = (5\mathbf{i} - 3\mathbf{j} - \mathbf{k})\,\text{N}$, $\mathbf{F}_3 = (6\mathbf{i} - \mathbf{j} + 3\mathbf{k})\,\text{N}$.

3 Four forces $\mathbf{F}_1$, $\mathbf{F}_2$, $\mathbf{F}_3$ and $\mathbf{F}_4$ act at points of a rigid body with position vectors $\mathbf{r}_1$, $\mathbf{r}_2$, $\mathbf{r}_3$ and $\mathbf{r}_4$ respectively relative to a fixed origin where

$\mathbf{F}_1 = (3\mathbf{i} - 4\mathbf{j} + 5\mathbf{k})\,\text{N}$ $\mathbf{r}_1 = (4\mathbf{i} - 2\mathbf{j} + 6\mathbf{k})\,\text{m}$
$\mathbf{F}_2 = (2\mathbf{i} + 6\mathbf{j} - 3\mathbf{k})\,\text{N}$ $\mathbf{r}_2 = (3\mathbf{i} + 8\mathbf{j} - 2\mathbf{k})\,\text{m}$
$\mathbf{F}_3 = (-6\mathbf{i} + 2\mathbf{j} + 2\mathbf{k})\,\text{N}$ $\mathbf{r}_3 = (7\mathbf{i} - \mathbf{k})\,\text{m}$
$\mathbf{F}_4 = (\mathbf{i} - 4\mathbf{j} - 4\mathbf{k})\,\text{N}$ $\mathbf{r}_4 = (6\mathbf{j} + 5\mathbf{k})\,\text{m}$

Show that the body is in equilibrium.

4 Forces $\mathbf{F}_1$, $\mathbf{F}_2$ and $\mathbf{F}_3$ act at points of a body with position vectors $\mathbf{r}_1$, $\mathbf{r}_2$ and $\mathbf{r}_3$ relative to a fixed origin, where

$\mathbf{F}_1 = (3\mathbf{i} - 3\mathbf{j} + 4\mathbf{k})\,\text{N}$ $\mathbf{r}_1 = (\mathbf{i} + \mathbf{j} + \mathbf{k})\,\text{m}$
$\mathbf{F}_2 = (3\mathbf{i} + 4\mathbf{j} + 3\mathbf{k})\,\text{N}$ $\mathbf{r}_2 = (3\mathbf{i} + 2\mathbf{j} + \mathbf{k})\,\text{m}$
$\mathbf{F}_3 = (-4\mathbf{i} - 2\mathbf{j} + \mathbf{k})\,\text{N}$ $\mathbf{r}_3 = (2\mathbf{i} - \mathbf{j})\,\text{m}$

(a) Reduce the system to a single force acting at the origin together with a couple.

(b) Reduce the system to a single force acting at the point with position vector $(\mathbf{i} - \mathbf{j} + 2\mathbf{k})\,\text{m}$ together with a couple.

(c) In each case, state in vector form an equation for the line of action of the force.

5 Forces $\mathbf{F}_1$, $\mathbf{F}_2$ and $\mathbf{F}_3$ act on a rigid body at points having position vectors $\mathbf{r}_1$, $\mathbf{r}_2$ and $\mathbf{r}_3$ relative to a fixed origin O where

$\mathbf{F}_1 = (4\mathbf{i} - 2\mathbf{j} + 2\mathbf{k})\,\text{N}$ $\mathbf{r}_1 = (\mathbf{i} + 4\mathbf{j} - 2\mathbf{k})\,\text{m}$
$\mathbf{F}_2 = (-2\mathbf{i} - 3\mathbf{j} - 6\mathbf{k})\,\text{N}$ $\mathbf{r}_2 = (2\mathbf{i} - 2\mathbf{j} - 2\mathbf{k})\,\text{m}$
$\mathbf{F}_3 = (-2\mathbf{i} + 5\mathbf{j} + 4\mathbf{k})\,\text{N}$ $\mathbf{r}_3 = (3\mathbf{i} + 2\mathbf{k})\,\text{m}$

Show that the system reduces to a couple.

6 Forces $\mathbf{F}_1$, $\mathbf{F}_2$ and $\mathbf{F}_3$ of magnitudes 14 N, 15 N and 9 N act on a body through the points with position vectors $(3\mathbf{i} + 4\mathbf{j} - 2\mathbf{k})$ m, $(2\mathbf{i} + 3\mathbf{j} + 4\mathbf{k})$ m and $(-2\mathbf{i} + \mathbf{j} + \mathbf{k})$ m.
$\mathbf{F}_1$ acts parallel to $2\mathbf{i} + 3\mathbf{j} + 6\mathbf{k}$.
$\mathbf{F}_2$ acts parallel to $3\mathbf{i} + 4\mathbf{j}$.
$\mathbf{F}_3$ acts parallel to $-8\mathbf{i} + 4\mathbf{j} - \mathbf{k}$.
Reduce the system to a single force $\mathbf{F}_4$ acting at the point with position vector $(2\mathbf{i} - 2\mathbf{j} + 4\mathbf{k})$ together with a couple.

7 Forces $\mathbf{F}_1$, $\mathbf{F}_2$ and $\mathbf{F}_3$ act on a rigid body at the points with position vectors $\mathbf{r}_1$, $\mathbf{r}_2$ and $\mathbf{r}_3$ relative to a fixed origin O where
$\mathbf{F}_1 = (3\mathbf{i} - 2\mathbf{j} + \mathbf{k})$ N $\mathbf{r}_1 = (7\mathbf{i} + 6\mathbf{j})$ m
$\mathbf{F}_2 = (4\mathbf{i} + 7\mathbf{j} + 2\mathbf{k})$ N $\mathbf{r}_2 = (-\mathbf{i} - 8\mathbf{j} - 4\mathbf{k})$ m
$\mathbf{F}_3 = (2\mathbf{i} - \mathbf{j} + 3\mathbf{k})$ N $\mathbf{r}_3 = (5\mathbf{i} + 7\mathbf{j} - 3\mathbf{k})$ m
Show that the three forces can be reduced to a single force $\mathbf{R}$.
Find a vector equation for the line of action of $\mathbf{R}$.

8 Four forces $(-\mathbf{i} - \mathbf{j} + \mathbf{k})$ N, $(2\mathbf{i} - \mathbf{k})$ N, $(\mathbf{i} + \mathbf{j} + \mathbf{k})$ N and $\mathbf{F}$ act on a rigid body at the points with position vectors $(-10\mathbf{i} + 2\mathbf{j})$ m, $6\mathbf{j}$ m, $3\mathbf{k}$ m and $(2\mathbf{i} + 3\mathbf{k})$ m relative to a fixed origin. The four forces are equivalent to a couple. Find
(a) the force $\mathbf{F}$,
(b) the magnitude of the couple $\mathbf{G}$,
stating your units clearly.

9 A rigid body is acted on by three forces:
$\mathbf{F}_1 = (4\mathbf{i} - 3\mathbf{j} + 2\mathbf{k})$ N at the point A with position vector $9\mathbf{i}$ m,
$\mathbf{F}_2 = (3\mathbf{i} + 2\mathbf{j} - 4\mathbf{k})$ N at the point B with position vector $(-\mathbf{i} - \mathbf{j} + 6\mathbf{k})$ m, and $\mathbf{F}_3$.

Given that the body is in equilibrium,
(a) find $\mathbf{F}_3$,
(b) show that the vector equation of the line of action of $\mathbf{F}_3$ can be written in the form

$$\mathbf{r} = [(5\mathbf{i} + 3\mathbf{j} - 2\mathbf{k}) + \lambda(-7\mathbf{i} + \mathbf{j} + 2\mathbf{k})]\text{ m}$$

where λ is a parameter.

10 Forces $\mathbf{F}_1$, $\mathbf{F}_2$ and $\mathbf{F}_3$ act on a rigid body at the points with position vectors $\mathbf{r}_1$, $\mathbf{r}_2$ and $\mathbf{r}_3$ relative to a fixed origin O where

$\mathbf{F}_1 = (p\mathbf{i} - \mathbf{j} + 3\mathbf{k}) \quad \mathbf{r}_1 = \mathbf{i} - 2\mathbf{j} - \mathbf{k}$

$\mathbf{F}_2 = (4\mathbf{i} + q\mathbf{j} + 5\mathbf{k}) \quad \mathbf{r}_2 = 7\mathbf{i} - 2\mathbf{j} + \mathbf{k}$

$\mathbf{F}_3 = (-6\mathbf{i} + r\mathbf{k}) \quad \mathbf{r}_3 = 3\mathbf{i} - 3\mathbf{j} + \mathbf{k}$

Given that this system of forces is equivalent to a couple, find the values of p, q and r and the moment of the couple.

11 Two forces $(\mathbf{i} + 3\mathbf{j} - 2\mathbf{k})$ N and $(\mathbf{i} - \mathbf{j} + \mathbf{k})$ N act through a point O of a rigid body. A couple of moment $(-2\frac{1}{2}\mathbf{i} + \mathbf{j} - 3\mathbf{k})$ N m also acts on the body. The system is equivalent to a single force $\mathbf{F}$. Find a vector equation for the line of action of $\mathbf{F}$.

SUMMARY OF KEY POINTS

1 To solve a vector differential equation of the type

$$\frac{d\mathbf{v}}{dt} = k\mathbf{v}$$

substitute $\mathbf{v} = u\mathbf{i} + w\mathbf{j}$ (2 dimensions)
or $\mathbf{v} = u\mathbf{i} + w\mathbf{j} + s\mathbf{k}$ (3 dimensions)

and solve the resulting scalar equations.

2 The vector differential equation

$$\frac{d^2\mathbf{r}}{dt^2} + 2k\frac{d\mathbf{r}}{dt} + (k^2 + n^2)\mathbf{r} = \mathbf{0}$$

should also be solved by substitution, as in key point 1.

3 The general solution of the differential equation $\frac{d^2\mathbf{r}}{dt^2} + 2k\frac{d\mathbf{r}}{dt} + (k^2 + n^2)\mathbf{r} = \mathbf{g}(t)$ is

$$\mathbf{r} = \text{complementary function} + \text{particular integral}$$

4 The vector differential equation $\frac{d\mathbf{r}}{dt} + f(t)\mathbf{r} = \mathbf{a}e^{bt}$ must first be multiplied by the integrating factor $e^{\int f(t)dt}$. It can then be integrated directly, without any substitution.

5 The work done by a constant force $\mathbf{F}$ which moves its point of application through a displacement $\mathbf{d}$ is given by

$$\text{work done} = \mathbf{F}.\mathbf{d}$$

6 The vector moment of the force **F** about the origin O is given by

$$\text{vector moment of force} = \mathbf{r} \times \mathbf{F}$$

where **r** is the position vector, relative to O, of any point on the line of action of **F**.

7 A system of forces $\mathbf{F}_1, \mathbf{F}_2, \ldots, \mathbf{F}_n$ with lines of action passing through the points with position vectors $\mathbf{r}_1, \mathbf{r}_2, \ldots, \mathbf{r}_n$ relative to an origin O, is equivalent to a single force

$$\mathbf{F} = \sum_{i=1}^{n} \mathbf{F}_i \text{ acting at } O$$

and a couple of moment

$$\mathbf{G} = \sum_{i=1}^{n} \mathbf{r}_i \times \mathbf{F}_i.$$

8 If the resultant force **F** is required to act at some point other than O, **F** is unchanged but the couple will have a different moment.

9 A system of forces is in equilibrium if $\mathbf{F} = \mathbf{0}$ and the sum of the moments of the forces about any point is zero.

10 A system of forces in three dimensions is either in equilibrium or can be reduced to a single force, a couple or a force and a couple.

Motion of a particle with varying mass

2

In all the problems considered thus far, the mass of the body or particle involved has been assumed to be constant. However, there are some problems where the mass of the body changes; for example, a rocket, whose mass becomes less as its motors burn up fuel; or a raindrop, whose mass becomes greater as it passes through a cloud. Such problems are usually called 'variable mass' problems, but in fact the mass in the problem is being redistributed rather than changing.

In problems of this kind, rather than write down the equations of motion of the system, we usually use the **impulse–momentum principle** (see Book M1, section 5.3). The system will be assumed to consist of the body in motion together with any matter that is ejected in a small interval of time δt or any matter that is absorbed in this time interval. At the beginning of this interval the various parts of the system will have their separate motions; at the end of the interval each part will normally have a different motion.

- **The impulse–momentum principle states: change of linear momentum of the whole system in time interval δt = impulse of external forces acting on whole system in time interval δt.**

Suppose the body picks up material of mass δm in the interval and that its speed increases by δv. The quantity $\delta m \cdot \delta v$ tends to zero more rapidly than δm or δv do as $\delta t \to 0$, and so such products may be neglected in writing down the impulse–momentum equation. You can then set up a differential equation by dividing by δt and taking the limit as $\delta t \to 0$.

In each case proceed from first principles and derive the differential equation from scratch. Do not try to quote standard results. The method is illustrated in the following examples.

Example 1

A body has mass m and speed v at time t and picks up matter when falling under constant gravity. Obtain the differential equation satisfied by m

(a) when the matter being picked up is not moving,
(b) when the matter being picked up has speed u.

(a) Consider the system as the body of mass m and the material of mass δm.

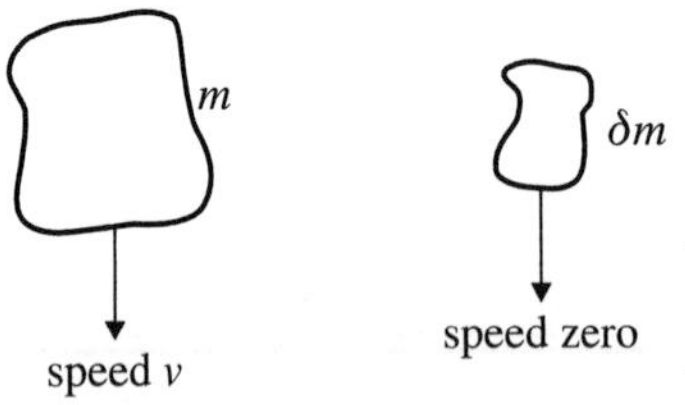

The initial momentum is then

$$mv + \delta m \times 0 = mv$$

At the end of the interval the body and material have coalesced to form a body of mass $(m + \delta m)$ moving with a speed $(v + \delta v)$. The final momentum is then

$$(m + \delta m)(v + \delta v)$$

The external force acting on the system is the total weight $mg + (\delta m)g$.

By the impulse–momentum principle:

$$\text{change in momentum} = \text{impulse of external force}$$

$$(m + \delta m)(v + \delta v) - mv = (m + \delta m)g\delta t \qquad (1)$$

so that $\quad mv + m\delta v + v\delta m + \delta m \cdot \delta v - mv = mg\delta t + g\delta m \cdot \delta t$

Neglecting the term $\delta m \cdot \delta v$ and dividing by δt gives

$$mg + g\delta m = m\frac{\delta v}{\delta t} + v\frac{\delta m}{\delta t}$$

Taking the limit as $\delta t \to 0$ and using the fact that $\delta m \to 0$ gives

$$mg = m\frac{\mathrm{d}v}{\mathrm{d}t} + v\frac{\mathrm{d}m}{\mathrm{d}t} = \frac{\mathrm{d}}{\mathrm{d}t}(mv) \qquad (2)$$

(b) If the matter picked up was moving with speed u rather than being at rest, the initial momentum would be $mv + \delta m \cdot u$ and equation (1) is then modified to

$$(m + \delta m)(v + \delta v) - (mv + \delta mu) = (m + \delta m)g\delta t$$

This leads to the differential equation

$$mg = m\frac{\mathrm{d}v}{\mathrm{d}t} + v\frac{\mathrm{d}m}{\mathrm{d}t} - u\frac{\mathrm{d}m}{\mathrm{d}t} \qquad (3)$$

Notice that equation (2) expresses Newton's second law in the form

(rate of change of downward momentum) = (downward force)

However, equation (3) shows that the result must be modified when the additional mass is not picked up from rest.

Example 2

A body moves vertically upwards under gravity. At time t its mass is m and its speed is v. The body ejects material at a rate of k units of mass per second vertically downwards with a speed u relative to the body. Obtain the differential equation satisfied by m.

At the start of the interval the body has mass m and speed v. At the end of the time interval δt it has mass $(m + \delta m)$. A mass

$$-\delta m = k\delta t$$

has been ejected. The speed is now $(v + \delta v)$. The ejected mass has a speed $(v - u)$ at the start of the interval and $(v + \delta v - u)$ at the end of the interval. The upward momentum of the ejected mass lies between $-\delta m(v - u)$ and $-\delta m(v + \delta v - u)$, and since we neglect terms in $\delta m \cdot \delta v$ this will be taken as $-\delta m(v - u)$. Finally, the external force acting on the whole system is mg. Using the impulse–momentum principle gives

$$[(m + \delta m)(v + \delta v) + (-\delta m)(v - u)] - mv = -mg\delta t$$

Neglecting the term in $\delta m \cdot \delta v$, this reduces to

$$m\delta v + u\delta m = -mg\delta t$$

Dividing by δt and taking the limit as $\delta t \to 0$ gives

$$m\frac{\mathrm{d}v}{\mathrm{d}t} + u\frac{\mathrm{d}m}{\mathrm{d}t} = -mg$$

but
$$\frac{\mathrm{d}m}{\mathrm{d}t} = -k$$

so
$$m\frac{\mathrm{d}v}{\mathrm{d}t} - ku = -mg$$

Example 3

A spherical hailstone, falling under gravity in still air, increases its radius by condensation according to the law $\frac{\mathrm{d}r}{\mathrm{d}t} = kr$, where k is constant. Neglecting air resistance, show that the hailstone approaches the limiting speed $\left(\frac{g}{3k}\right)$.

By the impulse–momentum principle:

$$\text{change in momentum} = \text{impulse of external force}$$

Since the added material is picked up from rest this gives

$$(m + \delta m)(v + \delta v) - mv = (m + \delta m)g\delta t$$

Neglecting the term $\delta m \cdot \delta v$, dividing by δt and taking the limit as $\delta t \to 0$ gives

$$m\frac{\mathrm{d}v}{\mathrm{d}t} + v\frac{\mathrm{d}m}{\mathrm{d}t} = mg$$

As the hailstone is spherical in this case,

$$m = \tfrac{4}{3}\pi r^3 \rho$$

where ρ is the density of the ice. The equation becomes

$$\tfrac{4}{3}\pi r^3 \rho \frac{dv}{dt} + v\frac{d}{dt}(\tfrac{4}{3}\pi r^3 \rho) = \tfrac{4}{3}\pi r^3 \rho g$$

or

$$r^3 \frac{dv}{dt} + v\frac{d}{dt}(r^3) = r^3 g$$

$$r^3 \frac{dv}{dt} + 3vr^2 \frac{dr}{dt} = r^3 g$$

Using $\dfrac{dr}{dt} = kr$ gives

$$r^3 \frac{dv}{dt} + 3kr^3 v = r^3 g$$

or

$$\frac{dv}{dt} = g - 3kv$$

You can find the limiting speed by setting $\dfrac{dv}{dt} = 0$.

So

$$v(\text{limiting}) = \left(\frac{g}{3k}\right)$$

Example 4

A particle P increases its mass, through condensation of moisture, at a constant time rate $\dfrac{m_0}{\tau}$, where m_0 is the initial mass of P and τ is a constant. The particle moves freely under gravity and is projected from O, the origin of coordinates, with a velocity whose horizontal and vertical components are U and V respectively. Show that the coordinates (x, y) of P, when its mass is m, are given by

$$x = U\tau \ln\left(\frac{m}{m_0}\right)$$

$$y = \tfrac{1}{4}g\tau^2\left(1 - \frac{m^2}{m_0^2}\right) + (V\tau + \tfrac{1}{2}g\tau^2)\ln\left(\frac{m}{m_0}\right)$$

Suppose that at time t the velocity of P has components u and v. Since the matter is being picked up from rest, from example 1 we obtain the equations

$$\frac{d}{dt}(mu) = 0 \qquad (1)$$

$$\frac{d}{dt}(mv) = -mg \qquad (2)$$

Equation (1) comes from the fact that there is no horizontal force acting on P.
Equation (2) comes from the fact that gravity acts vertically downwards and v is the resolute of the velocity vertically upwards.

In addition, the variation of mass equation is

$$\frac{dm}{dt} = \frac{m_0}{\tau} \qquad (3)$$

Integrating equation (1) with respect to t gives

$$mu = m_0 U$$

So

$$m\frac{dx}{dt} = m_0 U \qquad (4)$$

This equation contains 3 variables, m, x and t. Eliminate t by dividing equation (3) by equation (4) to obtain

$$\frac{1}{m}\frac{dm}{dx} = \frac{1}{U\tau}$$

Integrating this equation with respect to x gives

$$\ln m = \frac{x}{U\tau} + \ln m_0$$

so

$$x = U\tau \ln\left(\frac{m}{m_0}\right)$$

Integrating (3) with respect to t gives

$$m = \frac{m_0 t}{\tau} + m_0 = \frac{m_0}{\tau}(t + \tau) \qquad (5)$$

Using (5) in equation (2) to eliminate the variable m from the right-hand side of equation (2) gives

$$\frac{d}{dt}(mv) = -\frac{m_0 g}{\tau}(t + \tau)$$

and integrating with respect to t gives

$$mv = -\frac{m_0 g}{\tau}\int (t + \tau)\,dt$$

$$= -\frac{m_0 g}{\tau}\left(\frac{t^2}{2} + \tau t\right) + C$$

$$= -\frac{m_0 g}{\tau}\left(\frac{t^2}{2} + \tau t\right) + m_0 V \qquad (6)$$

An expression for y in terms of m is required, so write the right-hand side of equation (6) in terms of m:

$$mv = -\frac{m_0 g}{2\tau}(t^2 + 2\tau t) + m_0 V$$
$$= -\frac{m_0 g}{2\tau}[(t+\tau)^2 - \tau^2] + m_0 V$$

so
$$mv = m\frac{dy}{dt} = -g\frac{m^2\tau}{2m_0} + \frac{m_0 g\tau}{2} + m_0 V$$

and
$$\frac{dy}{dt} = -\frac{gm\tau}{2m_0} + \frac{m_0}{m}(V + \tfrac{1}{2}g\tau) \qquad (7)$$

Dividing equation (7) by equation (3) to eliminate t gives

$$\frac{dy}{dm} = -\frac{gm\tau^2}{2{m_0}^2} + \frac{\tau}{m}(V + \tfrac{1}{2}g\tau)$$

Integrating with respect to m gives

$$y = -\frac{g\tau^2}{2{m_0}^2}\frac{m^2}{2} + (V\tau + \tfrac{1}{2}g\tau^2)\ln m + K$$

Since $m = m_0$ when $y = 0$:

$$K = \tfrac{1}{4}g\tau^2 - (V\tau + \tfrac{1}{2}g\tau^2)\ln m_0$$

and

$$y = \frac{1}{4}g\tau^2\left(1 - \frac{m^2}{{m_0}^2}\right) + (V\tau + \tfrac{1}{2}g\tau^2)\ln\left(\frac{m}{m_0}\right).$$

Exercise 2A

1 A rocket has initial total mass M. It propels itself by ejecting mass at a constant rate k per unit time with speed u relative to the rocket. The rocket is launched from rest vertically upwards. Show that its speed after time t is

$$-u\ln\left(1 - \frac{kt}{M}\right) - gt$$

provided that $ku > Mg$. Explain why this condition is required.

2 A spherical raindrop of radius a falls from rest under gravity. It falls through a stationary cloud so that, because of condensation, its radius increases with time at a constant rate k. Find the distance fallen by the raindrop after time t.

3 A small body, of mass m_0, is projected vertically upwards in a cloud. Its initial speed is $(2gk)^{\frac{1}{2}}$. During its motion the body picks up moisture from the stationary cloud. Its mass at height x above the point of projection is $m_0(1 + \alpha x)$, where α is a positive constant. Show that the greatest height h satisfies the equation

$$(1 + \alpha h)^3 = (1 + 3k\alpha)$$

4 A body consists of equal masses M of inflammable and non-inflammable material. The body descends freely under gravity from rest. The combustible part burns at a constant rate of kM per second, where k is a constant. The burning material is ejected vertically upwards with constant speed u relative to the body, and air resistance may be neglected. Show, using momentum considerations, that

$$\frac{\mathrm{d}}{\mathrm{d}t}[(2 - kt)v] = k(u - v) + g(2 - kt)$$

where v is the speed of the body at time t. Hence show that the body descends a distance

$$\frac{g}{2k^2} + \frac{u}{k}(1 - \ln 2)$$

before all the inflammable material is burnt.

5 A rocket is fired vertically upwards with initial speed V and is propelled by ejecting material downwards at a constant rate and with a constant speed u relative to the rocket. After a time T the propellant material is exhausted and the rocket still rising. Neglecting air resistance, show that the maximum height h attained is given by

$$2gh = (u \ln \lambda)^2 - 2Vu \ln \lambda + V^2 + 2guT[1 + (1 - \lambda)^{-1} \ln \lambda]$$

where λ is the ratio of the mass of the rocket alone to the total initial mass of the rocket and propellant.

6 A spherical hailstone falls freely from rest under gravity. It initially has radius a. As it falls its volume increases through condensation at a rate equal to λ times its surface area, where λ is a constant.

(a) Show that after a time t its radius r is equal to $a + \lambda t$.

(b) Show that the velocity v at time t satisfies the differential equation

$$\frac{\mathrm{d}v}{\mathrm{d}t} = g - \frac{3\lambda v}{r}$$

(c) Hence show that

$$4\lambda v = g[a + \lambda t - a^4(a + \lambda t)^{-3}]$$

7 A rocket-driven car of total mass M loses mass at a constant rate μ per unit time at a constant ejection speed V relative to the car. If the total resistance to motion is kv when the speed is v,

(a) show that the acceleration of the car along a straight horizontal road is

$$\frac{\mu V - kv}{M - \mu t}$$

at time t from the start.

(b) Hence show that the speed from rest is

$$\frac{\mu V}{k}\left[1 - \left(1 - \frac{\mu t}{M}\right)^{\frac{k}{\mu}}\right]$$

[E]

SUMMARY OF KEY POINTS

1 To solve problems involving changing mass, apply the impulse–momentum principle:

change of linear momentum of the whole system in time interval δt = impulse of external forces acting on whole system in time interval δt.

Review exercise 1

1 At time t a particle P, of mass m, has position vector $\mathbf{p}$, given by

$$\mathbf{p} = (3a\cos\omega t)\mathbf{i} + (4a\sin\omega t)\mathbf{j}$$

where a and ω are positive constants. Find, in terms of m, a, ω and t,

(a) the kinetic energy of P,

(b) the magnitude of the force acting on P.

A second particle Q has position vector $\mathbf{q}$, given by

$$\mathbf{q} = (3a\sin\omega t)\mathbf{j} + (4a\cos\omega t)\mathbf{k}$$

Find $\mathbf{r}$, the position vector of P relative to Q.

Evaluate $\mathbf{r}.\mathbf{r}$ and hence, or otherwise, show that the greatest and least distances between P and Q are $5a$ and a respectively.

[E]

2 Solve the vector differential equation

$$\frac{\mathrm{d}^2\mathbf{r}}{\mathrm{d}\theta^2} = \frac{\mathrm{d}\mathbf{r}}{\mathrm{d}\theta}$$

given that $\frac{\mathrm{d}\mathbf{r}}{\mathrm{d}\theta} = \mathbf{i}$ and $\mathbf{r} = \mathbf{j}$ when $\theta = 0$. [E]

3 Find the general solution of the vector differential equation

$$\frac{\mathrm{d}^2\mathbf{r}}{\mathrm{d}\theta^2} - 2\frac{\mathrm{d}\mathbf{r}}{\mathrm{d}\theta} + 10\mathbf{r} = 0$$

Given that $\mathbf{r} = \mathbf{j}$ when $\theta = 0$, find $\mathbf{r}$ when $\theta = \pi$.

4 At time t, the velocity $\mathbf{v}$ of a particle P satisfies the vector differential equation

$$\frac{\mathrm{d}\mathbf{v}}{\mathrm{d}t} + \frac{3\mathbf{v}}{T} = 0$$

where T is a constant. At time $t = 0$, the position vector of P is $a(\mathbf{i} + 2\mathbf{j})$ and its velocity is $3a(\mathbf{i} - \mathbf{j})/T$.

Find the position vector of P at any time t. [E]

5 Integrate the vector equation

$$\frac{d^2\mathbf{r}}{dt^2} + n^2\mathbf{r} = 0$$

to find **r**, given that $\mathbf{r} = (\mathbf{i} + \mathbf{j})a$ and $\frac{d\mathbf{r}}{dt} = (\mathbf{i} - 2\mathbf{k})b$ when $t = 0$. [E]

6 Solve the differential equation

$$\frac{d\mathbf{r}}{dt} = 2\mathbf{r}$$

given that, when $t = 0$, $\mathbf{r}.\mathbf{j} = 0$ and $\mathbf{r} \times \mathbf{j} = \mathbf{i} + \mathbf{k}$. [E]

7 At time t the position vector **r** of the point P satisfies the differential equation

$$\frac{d^2\mathbf{r}}{dt^2} - 9\omega^2\mathbf{r} = \mathbf{i}a\omega^2 \sin 2\omega t$$

where ω and a are constants. When $t = 0$, P passes with velocity $\omega a\mathbf{i}$ through the point with position vector $a\mathbf{j}$. Find a vector equation of the locus of P. [E]

8 Find the work done by a constant force **F**, where

$$\mathbf{F} = (6\mathbf{i} + 2\mathbf{j} - \mathbf{k})\,\text{N}$$

when the point of application is moved from the point with position vector **a** to the point with position vector **b**, where

$$\mathbf{a} = (\mathbf{i} + \mathbf{j} + 4\mathbf{k})\,\text{m}$$

and $$\mathbf{b} = (3\mathbf{i} - 3\mathbf{j} + 3\mathbf{k})\,\text{m}$$ [E]

9 Find the work done by a force **F**, where

$$\mathbf{F} = (3\mathbf{i} + 5\mathbf{j} + 4\mathbf{k})\,\text{N}$$

when the point of application is moved from the point with position vector **a** to the point with position vector **b**, where

$$\mathbf{a} = (2\mathbf{i} + 3\mathbf{j} - 3\mathbf{k})\,\text{m} \quad \mathbf{b} = (5\mathbf{i} + \mathbf{j} - \mathbf{k})\,\text{m}$$ [E]

10 Find the work done by a force **F** where

$$\mathbf{F} = (16\mathbf{i} + 12\mathbf{j})\,\text{N}$$

when the point of application is moved from the point with position vector **a** to the point with position vector **b**, where

$$\mathbf{a} = (-2\mathbf{i} + \mathbf{j} + \mathbf{k})\,\text{m} \quad \text{and} \quad \mathbf{b} = (10\mathbf{i} + 5\mathbf{j} + 11\mathbf{k})\,\text{m}$$ [E]

11 A force $\mathbf{F}$ of magnitude 18 N acts along the direction of the vector $(7\mathbf{i} + 4\mathbf{j} + 4\mathbf{k})$. A bead moves along a smooth straight wire from the point A to the point B, where A and B have position vectors

$$(\mathbf{i} + 3\mathbf{j} - 2\mathbf{k})\,\text{m} \quad \text{and} \quad (3\mathbf{i} + 5\mathbf{j} - \mathbf{k})\,\text{m}$$

respectively, under the action of $\mathbf{F}$ and the reaction $\mathbf{R}$ of the wire only. Find $\mathbf{R}$ and the work done by $\mathbf{F}$ in the motion.

[E]

12 The line of action of a force $2\mathbf{i} - \mathbf{j} + 3\mathbf{k}$ passes through the origin, and the line of action of a second force $\mathbf{i} + 2\mathbf{j} - \mathbf{k}$ passes through the point $(-1, 2, -3)$. Reduce the two forces to a single force acting at the origin together with a couple.

[E]

13 Forces $4\mathbf{j}$ N, $3\mathbf{k}$ N act through the points with position vectors $(\mathbf{i} + \mathbf{k})$ m and $(\mathbf{i} + \mathbf{j})$ m respectively. A third force acts through the point with position vector $(\mathbf{j} + \mathbf{k})$ m and is such that the three forces are equivalent to a couple. Find the vector moment and the magnitude of this couple. [E]

14 The forces $(b\mathbf{j} + c\mathbf{k})$, $(c\mathbf{k} + a\mathbf{i})$ and $(a\mathbf{i} + b\mathbf{j})$ act respectively through the three points with position vectors $a\mathbf{i}$, $b\mathbf{j}$ and $c\mathbf{k}$, where a, b and c are constants. Show that the force system is equivalent to a single force through the origin, and find its magnitude. [E]

15 Forces $\mathbf{F}_1$ and $\mathbf{F}_2$, where

$$\mathbf{F}_1 = (3\mathbf{i} + 7\mathbf{j} + 5\mathbf{k})\,\text{N}, \ \mathbf{F}_2 = (2\mathbf{i} - 3\mathbf{j} + \mathbf{k})\,\text{N}$$

act respectively through points whose position vectors with respect to an origin O are $\mathbf{r}_1$ and $\mathbf{r}_2$, where

$$\mathbf{r}_1 = (4\mathbf{i} + 9\mathbf{j} + 8\mathbf{k})\,\text{m}, \ \mathbf{r}_2 = (-3\mathbf{i} + 8\mathbf{j} + \mathbf{k})\,\text{m}$$

Show that the lines of action of the forces $\mathbf{F}_1$ and $\mathbf{F}_2$ meet at a point, and find the position vector of this point of intersection. A third force, $\mathbf{F}_3$, where $\mathbf{F}_3 = (4\mathbf{i} + 2\mathbf{j} + 3\mathbf{k})$ N, acts through the point with position vector $(a\mathbf{i} + b\mathbf{j} + 3\mathbf{k})$ m. Given that the system of forces $\mathbf{F}_1$, $\mathbf{F}_2$ and $\mathbf{F}_3$ can be represented by a single force $\mathbf{F}$ acting through the origin, find the values of the constants a and b. [E]

16 A force $\mathbf{F}_1$, of magnitude 26 N, acts along the direction of the vector $(4\mathbf{i} - 3\mathbf{j} + 12\mathbf{k})$. Given that the line of action of $\mathbf{F}_1$ passes through the point which has position vector $(2\mathbf{i} + \mathbf{j} - \mathbf{k})$ m, find the moment of $\mathbf{F}_1$ about the origin O.
A bead moves along a smooth straight wire from the point A to the point B, where

$$\overrightarrow{OA} = (3\mathbf{i} - 2\mathbf{j} + \mathbf{k})\,\text{m}, \quad \overrightarrow{OB} = (5\mathbf{i} - 22\mathbf{j} + 2\mathbf{k})\,\text{m}$$

under the influence of $\mathbf{F}_1$ and the reaction of the wire only. Find the work done by $\mathbf{F}_1$ in this motion.
Two other forces $\mathbf{F}_2$ and $\mathbf{F}_3$, of magnitudes 7 N and 9 N, act along the directions of the vectors $(2\mathbf{i} - 6\mathbf{j} + 3\mathbf{k})$ and $(-4\mathbf{i} + 7\mathbf{j} - 4\mathbf{k})$ respectively. Given that both $\mathbf{F}_2$ and $\mathbf{F}_3$ pass through the point with position vector $(\mathbf{i} + \mathbf{j} + \mathbf{k})$ m, reduce the three forces $\mathbf{F}_1$, $\mathbf{F}_2$ and $\mathbf{F}_3$ to a single force at O and a couple.
[E]

17 Define the moment about a point O of a force $\mathbf{F}$ with point of application P such that $\overrightarrow{OP} = \mathbf{r}$. Show that $\mathbf{F}$ can be considered to act at any point on its line of action without changing the moment. Forces $\mathbf{F}_1$, $\mathbf{F}_2$, where $\mathbf{F}_1 = 2\lambda\mathbf{i}$, $\mathbf{F}_2 = -3\lambda\mathbf{j}$, act at the origin O and a force $\mathbf{F}_3$, where $\mathbf{F}_3 = \lambda\mathbf{i} - \lambda\mathbf{j}$, acts at the point with position vector $a(\mathbf{i} + \mathbf{j})$, where a and λ are positive constants. Find
(a) the magnitude of the single resultant force $\mathbf{R}$ of the system,
(b) the moment of the forces $\mathbf{F}_1$, $\mathbf{F}_2$ and $\mathbf{F}_3$ about O.
By assuming $\mathbf{R}$ to act at a point with position vector $a(x\mathbf{i} + y\mathbf{j})$ find the moment of $\mathbf{R}$ about O and hence deduce the equation of the line of action of $\mathbf{R}$. [E]

18 Forces $\mathbf{F}_1$, $\mathbf{F}_2$, $\mathbf{F}_3$, where

$$\mathbf{F}_1 = (3\mathbf{i} - \mathbf{j} + 2\mathbf{k})\,\text{N}$$
$$\mathbf{F}_2 = (-\mathbf{i} - 4\mathbf{j} + \mathbf{k})\,\text{N}$$
$$\mathbf{F}_3 = (\mathbf{i} + \mathbf{j} - 2\mathbf{k})\,\text{N}$$

act at points with position vectors $\mathbf{r}_1$, $\mathbf{r}_2$, $\mathbf{r}_3$, where

$$\mathbf{r}_1 = (6\mathbf{i} - \mathbf{j} + \mathbf{k})\,\text{m}$$
$$\mathbf{r}_2 = (\mathbf{i} - 8\mathbf{j} + \mathbf{k})\,\text{m}$$
$$\mathbf{r}_3 = (\mathbf{i} - 2\mathbf{j} + 3\mathbf{k})\,\text{m}$$

When a fourth force $\mathbf{F}_4$ is added to these three forces, the system is in equilibrium. Find $\mathbf{F}_4$ and a vector equation of its line of action.
Find also the moment of $\mathbf{F}_4$ about the origin O. [E]

19 A rigid body is acted on by a force $\mathbf{F}_1$ at the point with position vector $\mathbf{r}_1$ and by a force $\mathbf{F}_2$ at the point with position vector $\mathbf{r}_2$ where

$$\mathbf{F}_1 = (-7\mathbf{i} + \mathbf{j} + 2\mathbf{k})\,\text{N}$$
$$\mathbf{F}_2 = (4\mathbf{i} - 3\mathbf{j} + 2\mathbf{k})\,\text{N}$$
$$\mathbf{r}_1 = (-2\mathbf{i} + 4\mathbf{j})\,\text{m}$$
$$\mathbf{r}_2 = (9\mathbf{i})\,\text{m}$$

A third force $\mathbf{F}_3$ also acts on the body.
Given that the body is in equilibrium under the action of the three forces, find $\mathbf{F}_3$ and show that the vector equation of the line of action of $\mathbf{F}_3$ may be written in the form

$$\mathbf{r} = [(5\mathbf{i} + 3\mathbf{j} - 2\mathbf{k}) + s(3\mathbf{i} + 2\mathbf{j} - 4\mathbf{k})]\,\text{m}$$

where s is a parameter. [E]

20 A small bead of mass 0.4 kg is free to move along a smooth straight wire joining the points A and B, whose position vectors relative to a fixed origin are $(\mathbf{i} - 4\mathbf{j} + 2\mathbf{k})$ m and $(4\mathbf{i} - 2\mathbf{j} + 3\mathbf{k})$ m respectively. Apart from the normal contact force due to the wire, the only force acting upon the bead is $\mathbf{F}$, where $\mathbf{F}$ is constant and in the direction of the vector $(5\mathbf{i} + 3\mathbf{j} + 4\mathbf{k})$. The bead starts from rest at A and has a speed of $5\,\text{m}\,\text{s}^{-1}$ on reaching B.
Find the magnitude of $\mathbf{F}$. [E]

21 Three forces, $(-\mathbf{i} + \mathbf{j})$ newtons, $(2\mathbf{j} - 2\mathbf{k})$ newtons and $\mathbf{F}$ newtons, act on a rigid body at the three points with position vectors $-17\mathbf{i}$ metres, $\frac{9}{2}\mathbf{i}$ metres and $4\mathbf{j}$ metres respectively relative to the origin. The forces are equivalent to a couple $\mathbf{G}$.
Find

(a) the force $\mathbf{F}$,

(b) the magnitude, in N m, of the couple $\mathbf{G}$. [E]

22 A rigid body is acted upon by forces $\mathbf{F}_1$ and $\mathbf{F}_2$, given by

$$\mathbf{F}_1 = (\mathbf{i} + 3\mathbf{j} - 5\mathbf{k})\,\text{N}$$
$$\mathbf{F}_2 = (4\mathbf{i} - \mathbf{j} + h\mathbf{k})\,\text{N}$$

where h is a constant.
The lines of action of $\mathbf{F}_1$ and $\mathbf{F}_2$ pass through the points with position vectors $(2\mathbf{i} - \mathbf{k})$ m and $(-\mathbf{j} + 2\mathbf{k})$ m respectively, relative to a fixed origin O.
The forces $\mathbf{F}_1$ and $\mathbf{F}_2$ can be replaced by an equivalent force system consisting of a single force $\mathbf{F}$ acting through O together with a couple $\mathbf{G}$.
Given that $|\mathbf{F}| = 3\sqrt{5}$ N, and that $|\mathbf{F}_2| > |\mathbf{F}_1|$,
(a) determine the value of h,
(b) find $\mathbf{G}$. [E]

23 Points A and B have position vectors $(3\mathbf{i} + \mathbf{j})$ m and $(5\mathbf{i} + 4\mathbf{j} + 2\mathbf{k})$ m respectively. A particle moves from rest at the point A to the point B under the action of a constant force $\mathbf{F}$ newtons only. Given that the work done by the force in moving the bead from A to B is 34 N m, find $\mathbf{F}$. [E]

24 The velocity $\mathbf{v}$ m s^{-1} at time t seconds of a particle moving in a plane satisfies

$$\frac{d\mathbf{v}}{dt} = 6\mathbf{v}, \text{ where } \mathbf{v} = (4\mathbf{i} + 2\mathbf{j}) \text{ when } t = 0.$$

Given that the particle passes through the point with position vector $(\mathbf{i} + \mathbf{j})$ m when $t = 0$,
(a) find the position vector of the particle at time t seconds.
(b) Find, to 2 significant figures, the time t when the magnitude of the acceleration first exceeds 100 m s^{-2}. [E]

25 The equation of motion of a particle P, of mass 1 kg, with position vector $\mathbf{r}$ metres at time t seconds is

$$\frac{\mathrm{d}^2\mathbf{r}}{\mathrm{d}t^2} + 3\frac{\mathrm{d}\mathbf{r}}{\mathrm{d}t} + 2\mathbf{r} = \mathbf{0}$$

At $t = 0$ the particle is at the point with position vector $2\mathbf{j}$ metres and its velocity is $(\mathbf{i} + \mathbf{j})\,\mathrm{m\,s^{-1}}$. Find $\mathbf{r}$ in terms of t.

[E]

26 At time t seconds, the velocity $\mathbf{v}$ metres per second of a particle P satisfies the equation

$$\frac{\mathrm{d}\mathbf{v}}{\mathrm{d}t} - 4\mathbf{v} = \mathbf{0}$$

When $t = 0$, the position vector of P, relative to a fixed origin O, is $(2\mathbf{i} + \mathbf{j})$ m and its velocity is $(\mathbf{i} - 4\mathbf{j})\,\mathrm{m\,s^{-1}}$.
Determine the position vector of P, relative to O, when $t = \ln 2$.

[E]

27 A rocket of initial total mass M propels itself by ejecting mass at a constant rate μ per unit time with speed u relative to the rocket. If the rocket is at rest directed vertically upwards, show that it will not initially leave the ground unless $\mu u > Mg$. Assuming this condition to hold, show that the velocity of the rocket after time t is given by

$$-u\ln\left(1 - \frac{\mu t}{M}\right) - gt$$

Show also that when the mass of the rocket has been reduced to half the initial value, its height above the ground will be

$$\frac{uM}{2\mu}\left(1 - \ln 2 - \frac{Mg}{4\mu u}\right)$$

28 A raindrop falls through a stationary cloud. Its mass m increases by accretion uniformly with the distance x fallen, so that

$$m = m_0(1 + kx)$$

Given that its speed v is zero when $x = 0$, show that

$$v^2 = \frac{2g}{3k}\left[1 + kx - \frac{1}{(1 + kx)^2}\right]$$

29 A particle whose initial mass is m is projected vertically upwards at time $t = 0$ with speed gT, where T is a constant. At time t its speed is u and its mass has increased to $me^{\frac{t}{T}}$. If the added mass is at rest when it is acquired, show that

$$\frac{d}{dt}(mue^{t/T}) = -mge^{t/T}$$

Deduce that the mass of the particle at its highest point is $2m$. If, instead, the added mass is falling with constant speed gT when it is acquired, find the mass of the particle at its highest point. [E]

30 A raindrop is observed at time $t = 0$ when it has mass m and downward velocity u. As it falls under gravity its mass increases by condensation at a constant rate λ and a resisting force acts on it, proportional to its speed and equal to λv when the speed is v. Show that

$$\frac{d}{dt}(M^2 v) = M^2 g$$

where $M = m + \lambda t$, and find the speed of the raindrop at time t.

31 A particle falls from rest under gravity through a stationary cloud. The mass of the particle increases by accretion from the cloud at a rate which at any time is mkv, where m is the mass and v the speed of the particle, and k is a constant. Show that after the particle has fallen a distance x

$$kv^2 = g(1 - e^{-2kx})$$

and find the distance the particle has fallen after time t.

32 At time t, the mass of a rocket is $M(1 - kt)$, where M and k are constants. At time t, the rocket is moving with speed v vertically upwards near the Earth's surface against constant gravity. Burnt fuel is expelled vertically downwards at speed u relative to the rocket. Show that

$$(1 - kt)\frac{dv}{dt} = ku - g(1 - kt)$$

Given that $v = 0$ when $t = 0$, find v in terms of g, u, k and t.

[E]

33 Initially the total mass of a rocket is M, of which kM is the mass of the fuel. Starting from rest, the rocket gives itself a constant vertical acceleration of magnitude g by ejecting fuel with constant speed u relative to itself. If m denotes its remaining mass at time t, show that the rate of decrease of m with respect to t is $2mg/u$, and deduce that

$$m = Me^{-2gt/u}$$

Find in terms of M, u and k an expression for the kinetic energy of the rocket when the fuel is exhausted. Find the value of k for which is energy is a maximum.
(Assume that the height reached is sufficiently small for g to be considered constant.) [E]

Moments of inertia of a rigid body

3

In Book M3 the motion of a particle moving in a circle was studied. When a body cannot be modelled as a particle, the study of its rotational motion is more complicated.

3.1 What is a moment of inertia?

The **inertia** of a particle or larger body is a measure of its reluctance to move. When a body can be modelled as a particle, its mass determines its inertia. For a larger body which is rotating about a fixed axis the particle model is no longer appropriate, and its reluctance to move is determined by a quantity called its **moment of inertia**.

Consider a lamina which is rotating about a fixed axis which passes through a point O of the lamina and is perpendicular to the plane of the lamina.

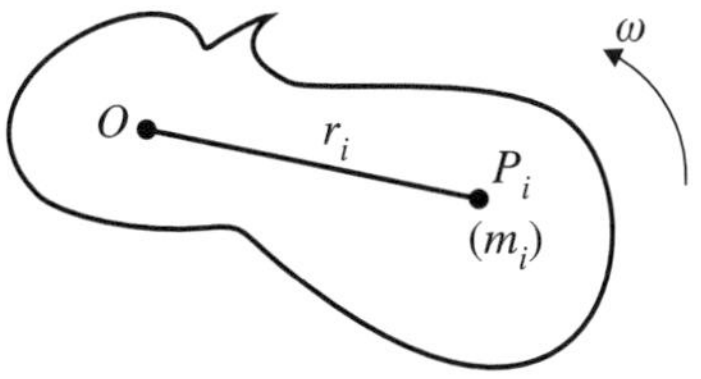

Think of the lamina as being composed of particles P_1, P_2, ..., P_n of masses m_1, m_2, ..., m_n which are at distances r_1, r_2, ..., r_n from the axis. Each of the radius vectors OP_1, OP_2, ..., OP_n will be rotating with the same angular speed ω, commonly referred to as the angular speed of the lamina. A typical particle P_i of mass m_i is shown in the diagram. This particle has a linear speed $r_i\omega$ (Book M3, chapter 4) and so its kinetic energy is

$$\tfrac{1}{2}m_i(r_i\omega)^2 = \tfrac{1}{2}m_i r_i{}^2\omega^2$$

The total kinetic energy (K.E.) of the lamina is obtained by summing over all the particles. This gives

$$\begin{aligned}\text{K.E. of lamina} &= \tfrac{1}{2}m_1r_1^2\omega^2 + \tfrac{1}{2}m_2r_2^2\omega^2 + \ldots + \tfrac{1}{2}m_nr_n^2\omega^2 \\ &= \tfrac{1}{2}\omega^2(m_1r_1^2 + m_2r_2^2 + \ldots + m_nr_n^2)\end{aligned}$$

The quantity $(m_1 r_1^2 + m_2 r_2^2 + m_n r_n^2)$ depends only on how the mass of the lamina is distributed relative to the axis of rotation. It is called the **moment of inertia** of the lamina about this axis.

If you consider the kinetic energy of a three-dimensional rotating body in the same way, you will get the same result.

- **The moment of inertia (M.I.) of a rigid body about an axis is $\sum_i m_i r_i^2$ where m_i is the mass of a typical particle and r_i is the distance of that particle from the axis.**

The symbol I is usually used for moment of inertia. When the mass of a body is measured in kilograms and distances in metres, the moment of inertia of the body is measured in kg m^2.

Note that since the moment of inertia of a body about an axis depends on the distances of the separate particles from that axis, it is meaningless to talk about the moment of inertia of a body without stating the axis of rotation.

The moment of inertia of a body about a given axis is a measure of its reluctance to turn about that axis. The larger the moment of inertia, the harder it is to change the angular speed of the body. The definition of the moment of inertia of a body about a given axis is the same for *all* bodies. However, in the following work we will only be considering objects which can be modelled as rigid bodies, as defined in Book M1, chapter 1.

3.2 Calculating moments of inertia

The method of calculating the moment of inertia of a body about a given axis is similar to the method of calculating the position of the centre of mass of a body (Book M3, chapter 5). Sometimes $\sum_i m_i r_i^2$ can be found by straightforward summation, but in other cases the summation will lead to integration.

Example 1

Find the moment of inertia of a circular hoop of mass M and radius a about an axis through its centre perpendicular to its plane.

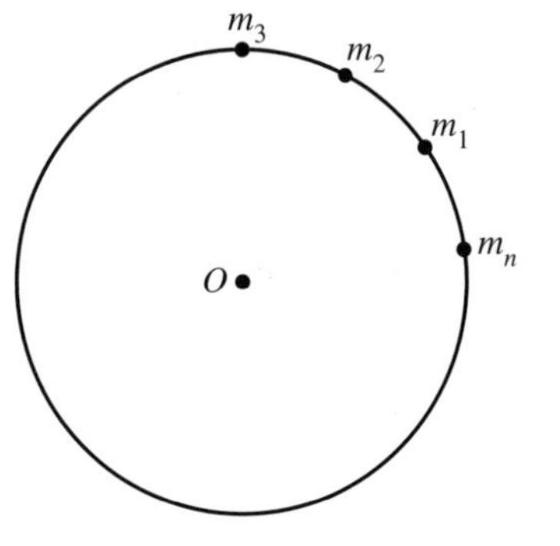

Think of the hoop as composed of a set of particles of masses $m_1, m_2, \ldots, m_n$ where $m_1 + m_2 + \ldots + m_n = M$, all at the same distance a from the centre O of the hoop and hence all at the same distance a from the axis.

So $\quad$ moment of inertia $= \sum_i m_i a^2$

$$= m_1 a^2 + m_2 a^2 + \ldots + m_n a^2$$
$$= a^2(m_1 + m_2 + \ldots + m_n)$$
$$= Ma^2$$

The particles do not have to be of equal mass and so the result does not depend on the hoop being uniform.

Example 2

Find the moment of inertia of a uniform rod of mass m and length $2a$ about an axis through its centre perpendicular to its length.

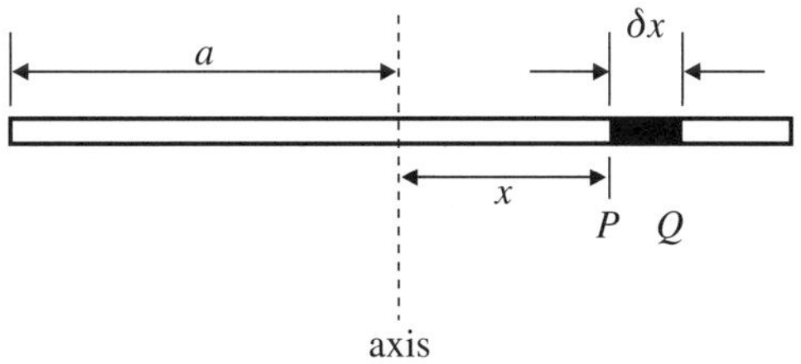

Consider the rod to be composed of small pieces. One such piece, PQ of length δx, at a distance x from the axis, is shown shaded in the diagram. As the rod has mass m and length $2a$ its mass per unit length is $\frac{m}{2a}$. So the mass of the small piece PQ is $\frac{m}{2a}\delta x$.

For this piece: $\quad mr^2 = \left(\frac{m}{2a}\delta x\right)x^2$

So for the whole rod:

$$\text{moment of inertia} = \sum mr^2$$
$$= \sum_{x=-a}^{x=a} \frac{mx^2}{2a}\delta x$$

If you let $\delta x \to 0$ then

$$\text{moment of inertia} = \lim_{\delta x \to 0} \sum_{x=-a}^{x=a} \frac{mx^2}{2a}\delta x$$
$$= \int_{-a}^{a} \frac{mx^2}{2a}\,\mathrm{d}x$$
$$= \left[\frac{mx^3}{3 \times 2a}\right]_{-a}^{a}$$
$$= \frac{ma^3}{6a} - \frac{m(-a)^3}{6a}$$
$$= \frac{ma^3}{6a} + \frac{ma^3}{6a}$$
$$= \tfrac{1}{3}ma^2$$

So the required moment of inertia is $\frac{1}{3}ma^2$.

Example 3

Find the moment of inertia of a uniform circular disc of mass m and radius r about an axis through its centre O perpendicular to its plane.

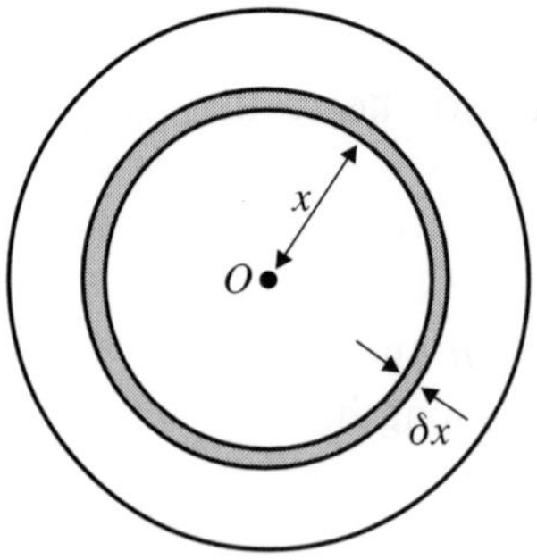

Consider the disc to be composed of a set of concentric hoops, centre O. Let a typical one have width δx and internal radius x, as shown shaded in the diagram.

The external radius of the hoop is $x + \delta x$. So

$$\begin{aligned}\text{area of the hoop} &= \pi(x + \delta x)^2 - \pi x^2 \\ &= 2\pi x \delta x + \pi(\delta x)^2\end{aligned}$$

As δx is small, $(\delta x)^2$ can be ignored compared with δx.

So $$\text{area of the hoop} = 2\pi x \delta x$$

The mass per unit area of the disc is $\dfrac{m}{\pi r^2}$.

So $$\begin{aligned}\text{mass of the hoop} &= \left(\frac{m}{\pi r^2}\right) 2\pi x \delta x \\ &= \frac{2mx}{r^2}\delta x\end{aligned}$$

Using the formula

$$\text{M.I. of a hoop} = \text{mass} \times \text{radius}^2$$

as found in example 1,

$$\begin{aligned}\text{M.I. of the typical hoop} &= \left(\frac{2mx}{r^2}\delta x\right) \times x^2 \\ &= \frac{2mx^3}{r^2}\delta x\end{aligned}$$

And so for the disc:

$$\text{M.I.} = \sum_{x=0}^{x=r} \frac{2mx^3}{r^2}\delta x$$

Letting $\delta x \to 0$:

$$\begin{aligned}\text{M.I.} &= \int_0^r \frac{2mx^3}{r^2}\,\mathrm{d}x \\ &= \left[\frac{2mx^4}{4r^2}\right]_0^r \\ &= \frac{2mr^4}{4r^2} \\ &= \tfrac{1}{2}mr^2\end{aligned}$$

3.3 The additive rule

You can deduce many moments of inertia from a few standard results by using various general rules. The first such rule is called the **additive rule**, and is readily obtained from the basic definition of moment of inertia.

Suppose two bodies have moments of inertia I_1 and I_2 about the same axis. The first body can be considered to be composed of particles of mass $m_1, m_2, \ldots, m_k$ at distances $r_1, r_2, \ldots, r_k$ from the axis and the second body to be composed of particles of mass $m_{k+1}, m_{k+2}, \ldots, m_n$ at distances $r_{k+1}, r_{k+2}, \ldots, r_n$ from the axis.

So $\quad I_1 = m_1r_1^2 + m_2r_2^2 + \ldots + m_k r_k^2$

and $\quad I_2 = m_{k+1}r_{k+1}^2 + m_{k+2}r_{k+2}^2 + \ldots + m_n r_n^2$

The moment of inertia of the composite body about this axis is

$$m_1r_1^2 + m_2r_2^2 + \ldots + m_kr_k^2 + m_{k+1}r_{k+1}^2 + \ldots + m_nr_n^2 = I_1 + I_2$$

Which gives us the **additive rule**:

- **If two bodies have moments of inertia I_1 and I_2 about the same axis then the M.I. of the composite body about the same axis is $I_1 + I_2$.**

Example 4

A uniform rod AB of length $2a$ and mass m has a particle of mass $2m$ attached to the point C of the rod where $BC = \frac{a}{2}$. Calculate the moment of inertia of the rod with the mass attached about an axis through the mid-point of AB and perpendicular to AB.

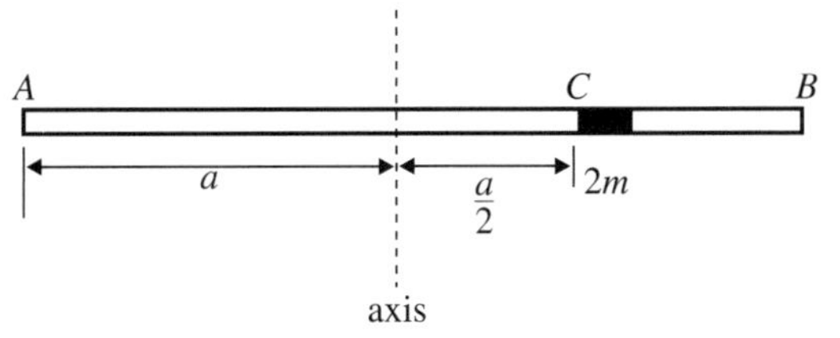

The standard result for the moment of inertia of a rod of length $2a$ and mass m about an axis through its mid-point perpendicular to its length is

$$\text{M.I. of rod} = \tfrac{1}{3}ma^2$$

(see example 2).

The moment of inertia of a particle of mass $2m$, distance $\frac{a}{2}$ from the axis, is

$$2m\left(\frac{a}{2}\right)^2 = \tfrac{1}{2}ma^2$$

By the additive rule:

$$\begin{aligned}\text{M.I. of (rod + particle)} &= \text{M.I. of rod} + \text{M.I. of particle} \\ &= \tfrac{1}{3}ma^2 + \tfrac{1}{2}ma^2 \\ &= ma^2(\tfrac{1}{3} + \tfrac{1}{2}) \\ &= \tfrac{5}{6}ma^2\end{aligned}$$

Example 5

A uniform ring of mass m is made from a circle of radius a, by cutting out a concentric circle of radius b. This ring, or **annulus**, D_1, is stuck to another uniform disc, D_2, of radius a and mass $2m$ so that the outer circumferences coincide. Show that the moment of inertia of the resulting flywheel about an axis perpendicular to the flywheel and through the centre O of the circles is $\frac{1}{2}m(3a^2 + b^2)$.

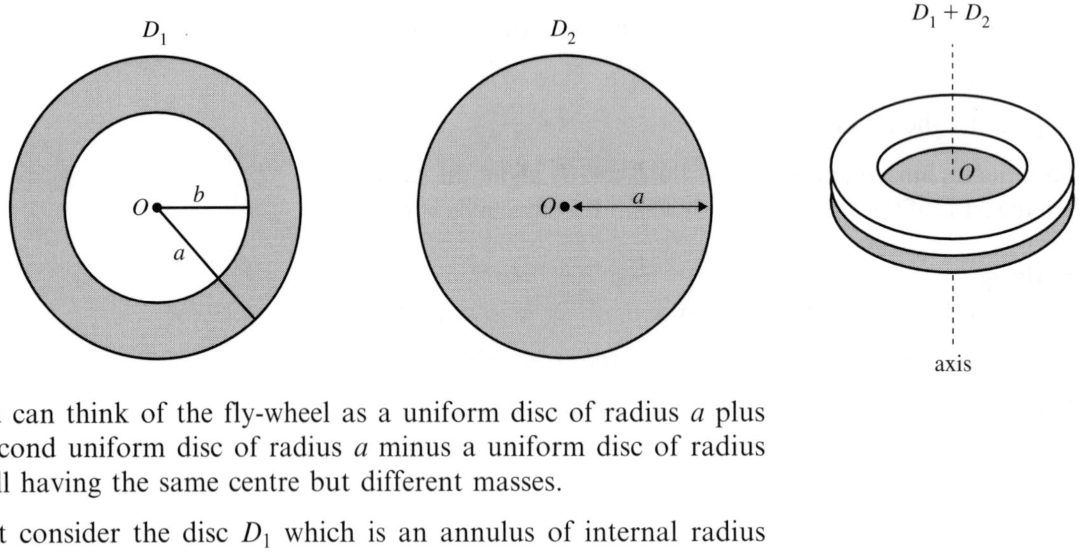

You can think of the fly-wheel as a uniform disc of radius a plus a second uniform disc of radius a minus a uniform disc of radius b, all having the same centre but different masses.

First consider the disc D_1 which is an annulus of internal radius b, external radius a and mass m.

$$\text{Area of annulus } D_1 = \pi a^2 - \pi b^2 = \pi(a^2 - b^2)$$

So $$\text{mass per unit area of } D_1 = \frac{m}{\pi(a^2 - b^2)}$$

So the mass of the circular disc, radius a, from which D_1 was cut is

$$\frac{\pi a^2 m}{\pi(a^2 - b^2)} = \frac{ma^2}{a^2 - b^2}$$

The moment of inertia of this circular disc of radius a about a perpendicular axis through its centre is

$$\begin{aligned}
&\tfrac{1}{2} \times \text{mass} \times (\text{radius})^2 \\
&= \tfrac{1}{2}\frac{ma^2}{a^2 - b^2} \times a^2 \\
&= \tfrac{1}{2}\frac{ma^4}{a^2 - b^2}
\end{aligned}$$

Similarly, the moment of inertia of a circular disc of radius b about a perpendicular axis through its centre is $\frac{1}{2}\frac{mb^4}{a^2 - b^2}$.

By the additive rule:

M.I. of annulus D_1 about a perpendicular axis through its centre

$$\begin{aligned}
&= \text{M.I. of circular disc radius } a - \text{M.I. of circular disc radius } b\text{, both about the same axis} \\
&= \tfrac{1}{2}\frac{ma^4}{a^2 - b^2} - \tfrac{1}{2}\frac{mb^4}{a^2 - b^2} \\
&= \tfrac{1}{2}m\frac{(a^4 - b^4)}{a^2 - b^2} \\
&= \tfrac{1}{2}m\frac{(a^2 - b^2)(a^2 + b^2)}{a^2 - b^2} \\
&= \tfrac{1}{2}m(a^2 + b^2)
\end{aligned}$$

M.I. of disc D_2, radius a, mass $2m$ about a perpendicular axis through its centre $= \frac{1}{2} \times 2ma^2 = ma^2$.

By the additive rule:

M.I. of flywheel about given axis

$$\begin{aligned}
&= \text{M.I. of annulus } D_1 + \text{M.I. of disc } D_2 \text{ about given axis} \\
&= \tfrac{1}{2}m(a^2 + b^2) + ma^2 \\
&= \tfrac{1}{2}m(3a^2 + b^2)
\end{aligned}$$

3.4 Using standard results

There are certain standard results for moments of inertia that you are given in the formula book for the examination. You can use these formulae, without proof, to work out other moments of inertia. However, you must be able to obtain them from first principles if requested to do so. Here are the three standard results which were derived in Examples 1, 2 and 3:

	Body	Axis	M.I.
■	**Uniform rod, mass m, length $2a$**	**through centre, perpendicular to rod**	$\frac{1}{3}ma^2$
■	**Circular hoop, mass m, radius r**	**through centre, perpendicular to plane of hoop**	mr^2
■	**Uniform circular disc, mass m, radius r**	**through centre, perpendicular to disc**	$\frac{1}{2}mr^2$

3.5 The stretching rule

The moment of inertia of a uniform rectangular lamina

Suppose you want to calculate the moment of inertia of a uniform rectangular lamina of mass m and sides $2a$ and $2b$ about an axis joining the mid-points of the sides of length $2a$. Think of the lamina as being composed of strips.

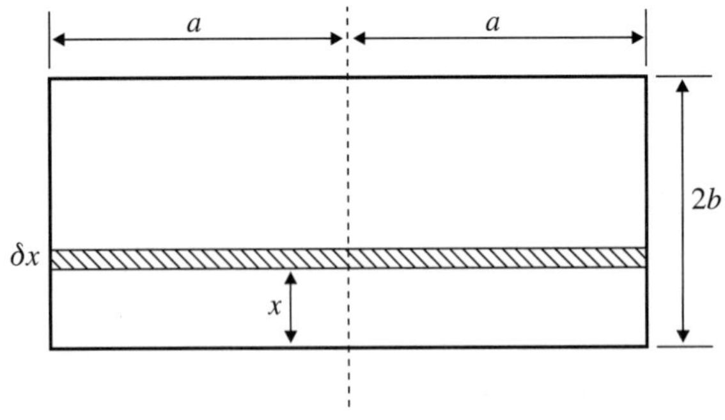

Let a typical strip have width δx and be distance x from a side of length $2a$ of the rectangle. This strip is shown shaded in the diagram.

$$\text{Mass per unit area of lamina} = \frac{m}{2a \times 2b} = \frac{m}{4ab}$$

So $$\text{mass of strip} = 2a\delta x \times \frac{m}{4ab} = \frac{m}{2b}\delta x$$

The moment of inertia of the strip about the given axis is the moment of inertia of a rod of length $2a$ about an axis through its centre perpendicular to its length. So

$$\text{M.I.} = \tfrac{1}{3} \times \text{mass} \times a^2 = \tfrac{1}{3}\left(\frac{m}{2b}\delta x\right)a^2 = \tfrac{1}{6}\frac{ma^2}{b}\delta x$$

Adding all the strips together and letting $\delta x \to 0$ gives

$$\text{M.I. of rectangle about given axis} = \int_0^{2b} \frac{ma^2}{6b} \, dx$$

$$= \frac{ma^2}{6b} \int_0^{2b} dx$$

$$= \frac{ma^2}{6b} \Big[x \Big]_0^{2b}$$

$$= \frac{ma^2}{6b} \times 2b$$

$$= \tfrac{1}{3} ma^2$$

So the moment of inertia of the rectangle is the same as the moment of inertia of a rod of the same mass and length $2a$ about an axis through its centre perpendicular to its length. This is an example of the **stretching rule**:

- **If one body can be obtained from another body by 'stretching' parallel to the axis without altering the distribution of mass *relative to the axis* then the moments of inertia of the two bodies about the axis are the same.**

The moment of inertia of a hollow cylinder

Another body which can be obtained by stretching is a cylinder.

Example 6

A hollow uniform circular cylinder of mass m and base radius a is open at both ends. Calculate the moment of inertia of the cylinder about its axis.

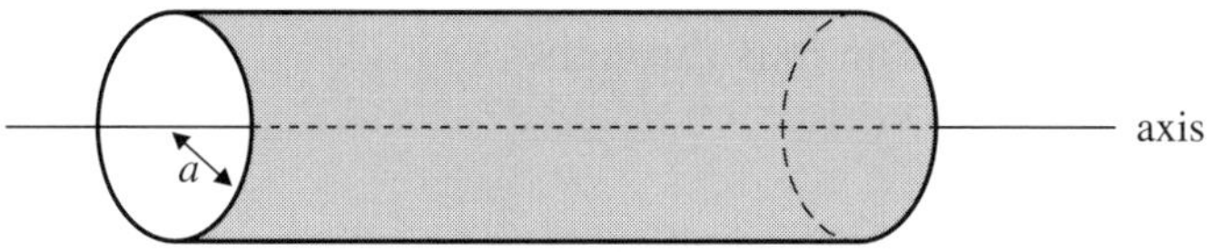

You can think of the cylinder as a 'stretch' of a circular hoop of mass m and radius a along an axis through the centre of the hoop and perpendicular to the plane of the hoop.

Thus M.I. of cylinder about its axis

= M.I. of hoop about perpendicular axis through the centre

$= ma^2$

The same method can be used to calculate the moment of inertia of a solid cylinder about its axis.

3.6 Radius of gyration

Suppose a body of mass m has a moment of inertia I about a given axis of rotation. A particle also of mass m placed at a distance k from the same axis will have moment of inertia mk^2 about that axis. The value of k which makes these two moments of inertia equal is called the **radius of gyration** of the original body about this axis.

The radius of gyration k is given by

$$I = mk^2$$

or

$$k = \sqrt{\left(\frac{I}{m}\right)}$$

Example 7

Calculate the radius of gyration of a rod of mass m and length $2a$ about an axis through its centre, perpendicular to its length.

$$\text{M.I. of rod about given axis} = \tfrac{1}{3}ma^2$$

Let the radius of gyration be k.

Then

$$k = \sqrt{\left(\frac{I}{m}\right)} = \sqrt{(\tfrac{1}{3}a^2)}$$

so

$$k = \frac{a\sqrt{3}}{3}$$

Exercise 3A

In questions 1–5, use integration to calculate the moments of inertia and radii of gyration of the bodies about the given axes.

1 A uniform rod of mass m and length $2a$ about an axis through one end, perpendicular to its length.

2 A uniform rod of mass m and length l about an axis perpendicular to its length through a point distance a from one end.

3 A uniform rod of mass m and length l about an axis through one end inclined at an angle θ to the rod.

4 A uniform triangular lamina of mass m in the shape of an isosceles triangle with base $2b$ and height h, about its axis of symmetry.

5 A uniform lamina of mass m bounded by the curve with equation $y^2 = 4ax$ and the line $x = 4a$ about the x-axis.

In questions 6–10 use standard results (see the end of this chapter) to calculate the required moments of inertia.

6 A circular hoop of mass m and radius r, with three particles of masses m, $2m$ and $3m$ attached to points on the hoop, about an axis through its centre.

7 A uniform rod of length l and mass m, about an axis through one end perpendicular to its length. (Hint: consider the rod to be half of a rod of length $2l$ and mass $2m$.)

8 A uniform rectangular lamina of mass m and sides l and b, about an axis through the mid-points of the sides of length l.

9 A uniform rod of mass m and length $2a$ with a particle of mass $2m$ attached to a point distance $\frac{a}{3}$ from one end, about an axis through its mid-point perpendicular to its length.

10 A uniform circular disc mass m and diameter d with particles of masses m, $2m$ and $4m$ attached at points distance $\frac{d}{3}$, $\frac{d}{3}$ and $\frac{d}{6}$ respectively from the centre, about a perpendicular axis through its centre.

11 An equilateral triangle ABC is formed by joining three light rods of length l. Particles of masses m, $2m$ and $3m$ are attached to the triangle at the mid-points of AB, BC and AC respectively. Calculate the moment of inertia of the system about an axis through A perpendicular to the plane of the triangle.

12 Calculate the moment of inertia of a uniform, hollow, closed cylinder of mass M, base radius r and height h, about its axis.

13 A uniform lamina of mass m is formed from a square lamina of side $2a$ by cutting away a square of side $2b$. Both squares have the same centre and their sides are parallel, as shown. A and B are the mid-points of opposite sides of the lamina. Calculate the moment of inertia of the lamina about AB.

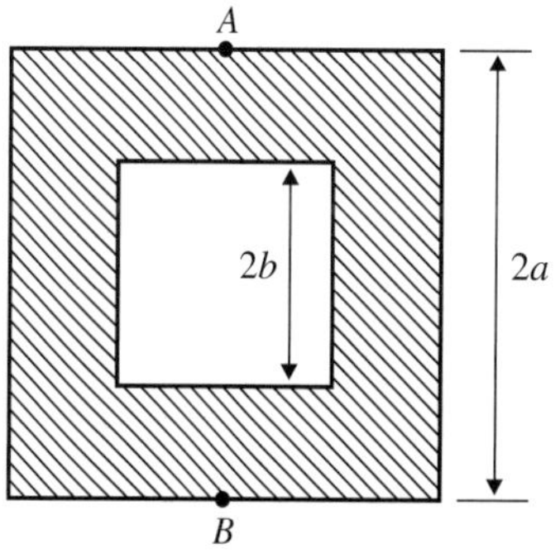

14 Calculate the moment of inertia of a uniform solid cylinder of mass M, base radius r and height h about its axis.

15 A hollow closed cylindrical container is made from thin sheets of metal. The height is H and the base radius is R. The mass per unit area of the metal used for the top and bottom is ρ and the mass per unit area of the metal used for the curved surface is 2ρ. Find, in terms of H, R and ρ, the moment of inertia of the cylinder about its axis of symmetry.

16 A composite body is made by joining two identical circular discs of radius r and mass m by means of a uniform rod. The rod is fixed to the centres of the two discs and is perpendicular to the plane of each disc. Find the moment of inertia of the body about an axis along the rod.

17 A particle of mass m is attached to the end P of a uniform rod PQ of length $6a$ and mass $5m$. Find the moment of inertia of the loaded rod about an axis through the centre of the rod which is perpendicular to the rod.

18 A fly-wheel is formed from a circular hoop of radius r and mass $4m$ and three uniform rods each of length $2r$ and mass m, which are fixed inside the hoop as shown in the diagram. The angles between the rods are all equal. Find the moment of inertia of the fly-wheel about an axis through its centre, perpendicular to the plane of the body.

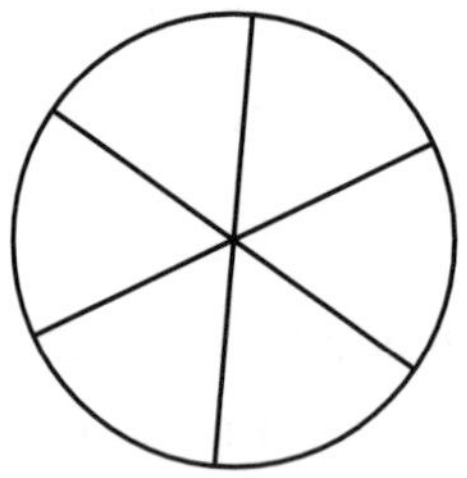

3.7 Moments of inertia of spheres

Using the additive rule and integration allows you to calculate the moments of inertia of more complicated bodies.

Moment of inertia of a solid sphere

Example 8

Find the moment of inertia of a uniform solid sphere of mass m and radius r about an axis coinciding with a diameter.

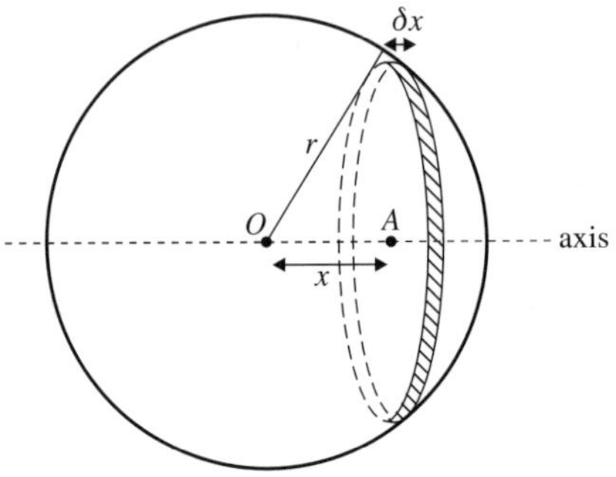

Consider the sphere to be composed of discs whose planes are perpendicular to the axis. Let a typical disc have thickness δx, be distance x from the centre O of the sphere and have centre A, as shown shaded in the diagram. By Pythagoras' theorem, the radius of the disc is $\sqrt{(r^2 - x^2)}$.

So $$\text{volume of disc} = \pi(r^2 - x^2)\delta x$$

The mass per unit volume of the sphere is

$$\frac{m}{\frac{4}{3}\pi r^3} = \frac{3m}{4\pi r^3}$$

So $$\text{mass of disc} = \pi(r^2 - x^2)\delta x \times \frac{3m}{4\pi r^3} = \frac{3m(r^2 - x^2)\delta x}{4r^3}$$

The axis passes through the centre of the disc and is perpendicular to the disc.

So $$\text{M.I. of disc} = \tfrac{1}{2} \times \text{mass} \times \text{radius}^2 = \tfrac{1}{2} \times \frac{3m(r^2 - x^2)}{4r^3} \times (r^2 - x^2)\delta x = \frac{3m}{8r^3}(r^2 - x^2)^2 \delta x$$

Adding the moments of inertia of the separate discs and letting $\delta x \rightarrow 0$ gives

$$\begin{aligned}
\text{M.I. of sphere about given axis} &= \int_{-r}^{r} \frac{3m}{8r^3}(r^2 - x^2)^2 \,dx \\
&= \frac{3m}{8r^3}\int_{-r}^{r} (r^4 - 2r^2x^2 + x^4)\,dx \\
&= \frac{3m}{8r^3}\left[r^4x - \frac{2r^2x^3}{3} + \frac{x^5}{5}\right]_{-r}^{r} \\
&= \frac{3m}{8r^3}\left[r^5 - \frac{2r^5}{3} + \frac{r^5}{5} - \left(-r^5 + \frac{2r^5}{3} - \frac{r^5}{5}\right)\right] \\
&= \frac{3m}{8r^3} \times 2 \times \frac{8r^5}{15} \\
&= \tfrac{2}{5}mr^2
\end{aligned}$$

- **So the moment of inertia of a uniform solid sphere of mass m and radius r about an axis coinciding with a diameter is $\frac{2}{5}mr^2$.**

The moment of inertia of a solid cone about its axis can be found in the same way.

Moment of inertia of a hollow sphere

To calculate the moment of inertia of a hollow sphere about an axis which coincides with a diameter, think of the sphere as a set of hoops. However, if you use slightly different radii for the two circles bounding each hoop, the surface of each hoop is slightly sloped and fits more closely to the actual surface of the sphere. This provides a closer approximation to the surface area, and hence the mass, of the sphere than can be obtained using 'straight'-sided hoops. Let the sphere have mass m and radius r.

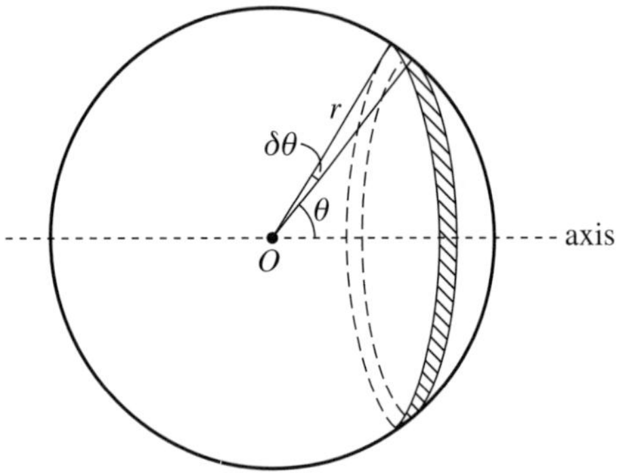

A typical hoop with planes perpendicular to the axis is shown in this diagram.

Let the angle between the axis and the radius of the sphere which joins a point on the outer circular boundary of the hoop to the centre of the sphere be θ, as shown. The angle between the axis and the radius of the sphere which joins a point on the inner circular boundary will be $\theta + \delta\theta$.

Each hoop is approximately a hollow cylinder. The centre of the hoop is on the axis of rotation, its radius is $r \sin\theta$ and its width is $r\delta\theta$.

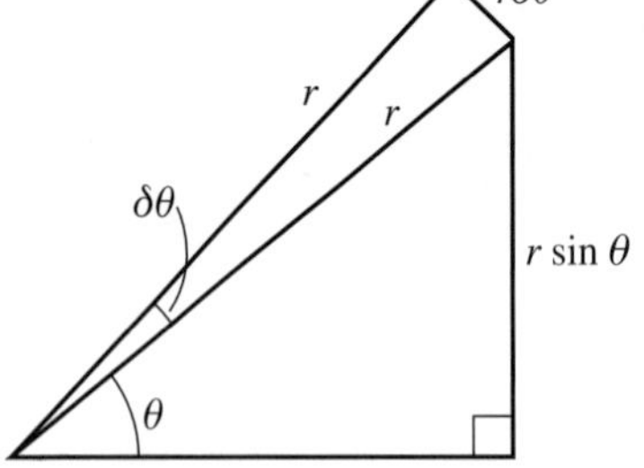

So $\qquad$ surface area of hoop $= 2\pi r \sin\theta \times r\delta\theta$

Surface area of sphere $= 4\pi r^2$.

So $\qquad$ mass per unit area of sphere $= \dfrac{m}{4\pi r^2}$

And so $\quad \text{mass of hoop} = 2\pi r^2 \sin\theta \times \dfrac{m}{4\pi r^2}\delta\theta$

$$= \frac{m\sin\theta}{2}\delta\theta$$

M.I. of hoop about perpendicular axis through centre

$$= \text{mass} \times \text{radius}^2$$

$$= \frac{m\sin\theta}{2} r^2 \sin^2\theta\delta\theta$$

Adding the moment of inertia of the separate hoops and letting $\delta\theta \to 0$ gives

$$\text{M.I. of sphere about a diameter} = \int_0^\pi \frac{mr^2}{2}\sin\theta . \sin^2\theta \, \mathrm{d}\theta$$

$$= \frac{mr^2}{2}\int_0^\pi \sin\theta(1 - \cos^2\theta)\,\mathrm{d}\theta$$

using $\sin^2\theta + \cos^2\theta \equiv 1$

$$= \frac{mr^2}{2}\int_0^\pi (\sin\theta - \sin\theta\cos^2\theta)\,\mathrm{d}\theta$$

$$= \frac{mr^2}{2}\left[-\cos\theta + \tfrac{1}{3}\cos^3\theta\right]_0^\pi$$

$$= \frac{mr^2}{2}[-(-1) + \tfrac{1}{3}(-1)^3 - (-1 + \tfrac{1}{3})]$$

$$= \frac{mr^2}{2}[1 - \tfrac{1}{3} + 1 - \tfrac{1}{3}]$$

$$= \frac{mr^2}{2} \times \tfrac{4}{3}$$

$$= \frac{2mr^2}{3}$$

- **So the moment of inertia of a hollow sphere of mass m and radius r about a diameter is $\dfrac{2mr^2}{3}$.**

3.8 The parallel axis theorem

Most of the moments of inertia found so far have been about an axis which passes through the centre of mass of the body. If you know the moment of inertia of a body about such an axis, you can calculate the moment of inertia of that body about *any* parallel axis by using the **parallel axis theorem**:

- **If the moment of inertia of a body of mass m about an axis through its centre of mass, G, is I_G then the moment of inertia about any axis parallel to the original axis and distance d from it is $I_G + md^2$.**

To prove this theorem, consider a body of mass M and let AB be the axis through the centre of mass G. Let $A'B'$ be an axis parallel to AB and distance d from AB.

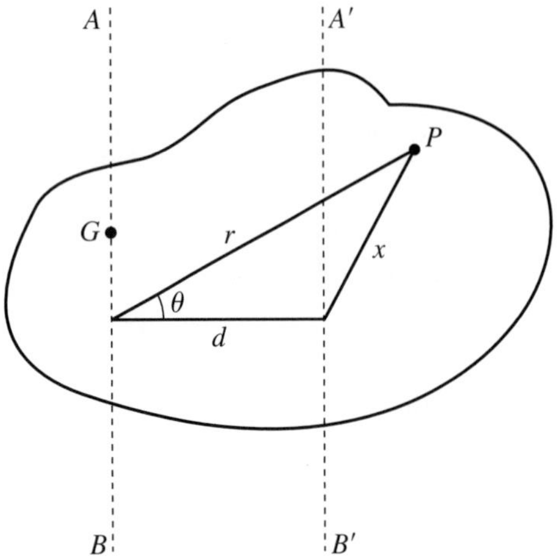

Consider a particle P of mass m which is a distance x from $A'B'$ and a distance r from AB.

By the cosine rule:

$$x^2 = d^2 + r^2 - 2dr\cos\theta$$

and so the moment of inertia of P about $A'B'$ is

$$mx^2 = m(d^2 + r^2 - 2dr\cos\theta)$$

So the moment of inertia of the body about $A'B'$ is

$$\begin{aligned}\Sigma mx^2 &= \Sigma m(d^2 + r^2 - 2dr\cos\theta)\\ &= d^2\Sigma m + \Sigma mr^2 - 2d\Sigma mr\cos\theta\end{aligned}$$

Now $\qquad \Sigma m = M \quad \text{and} \quad \Sigma mr^2 = I_G$

but $\qquad mr\cos\theta = \text{moment of mass of } P \text{ about } AB$

and since AB passes through the centre of mass of the body

$$\Sigma mr\cos\theta = 0$$

So the moment of inertia of the body about $A'B'$ is

$$I_G + Md^2$$

Example 9

Find the moment of inertia of a uniform hoop of radius a and mass m about an axis through a point A of the hoop perpendicular to the plane of the hoop.

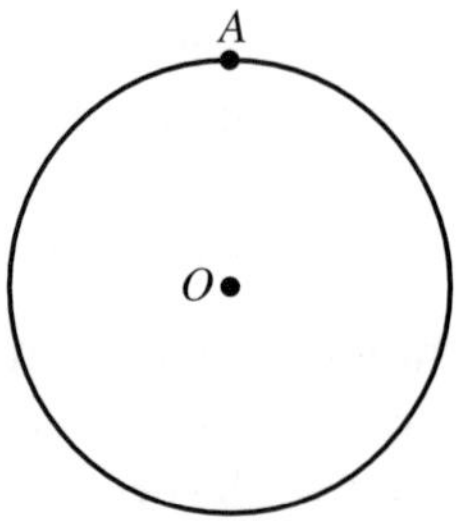

The moment of inertia of a hoop about an axis through its centre O perpendicular to the plane of the hoop is ma^2. Let A be any point on the hoop. Then $OA = a$ and by the parallel axis theorem:

M.I. of hoop about axis through A perpendicular to plane of hoop

$$= \text{M.I. of hoop about parallel axis through } O + ma^2$$
$$= ma^2 + ma^2$$
$$= 2ma^2$$

If the moment of inertia of a body about *any axis* is given, then you can calculate the moment of inertia about *any other parallel axis*, as long as you know the distances of each axis from the centre of mass. The calculation must be performed in two stages, as shown in the next example.

Example 10

A uniform equilateral triangular lamina ABC of mass m has $AB = AC = BC = l$. The moment of inertia of the lamina about an axis through A perpendicular to its plane is $\frac{5}{12}ml^2$. Calculate the moment of inertia of the lamina about a parallel axis through D, the mid-point of BC.

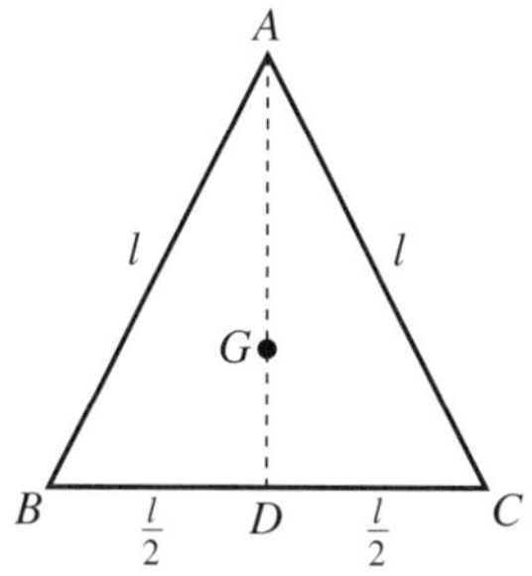

The centre of mass of the lamina is at G, where $GD = \frac{1}{3}AD$ (Book M2, section 2.2). Hence by Pythagoras' theorem in $\triangle ACD$,

$$GD = \tfrac{1}{3}\sqrt{\left(l^2 - \frac{l^2}{4}\right)} = \tfrac{1}{3}\sqrt{\left(\frac{3l^2}{4}\right)} = \frac{l\sqrt{3}}{6}$$

and so $$AG = 2GD = \frac{l\sqrt{3}}{3}$$

By the parallel axis theorem:

$$\text{M.I. about given axis through } A =$$
$$\text{M.I. about parallel axis through } G + [m \times (AG)^2]$$

So $\quad$ M.I. about axis through $G = \frac{5}{12}ml^2 - \left(m \times \frac{3l^2}{9}\right)$

$$= \tfrac{5}{12}ml^2 - \tfrac{1}{3}ml^2 = \tfrac{1}{12}ml^2$$

M.I. about parallel axis through D

$$= \text{M.I. about axis through } G + [m \times (GD)^2]$$

$$= \tfrac{1}{12}ml^2 + \left(m \times \frac{3l^2}{36}\right)$$

$$= \tfrac{1}{6}ml^2$$

3.9 The perpendicular axes theorem for a lamina

When you know the moments of inertia of a lamina about two perpendicular axes Ox and Oy in the plane of the lamina, you can find the moment of inertia of the lamina about the axis through O perpendicular to the plane of the lamina by adding these two moments of inertia.

Consider a plane lamina which is composed of particles P_i of mass m_i situated at points with coordinates (x_i, y_i) relative to the axes Ox and Oy. The axes are fixed in the plane of the lamina.

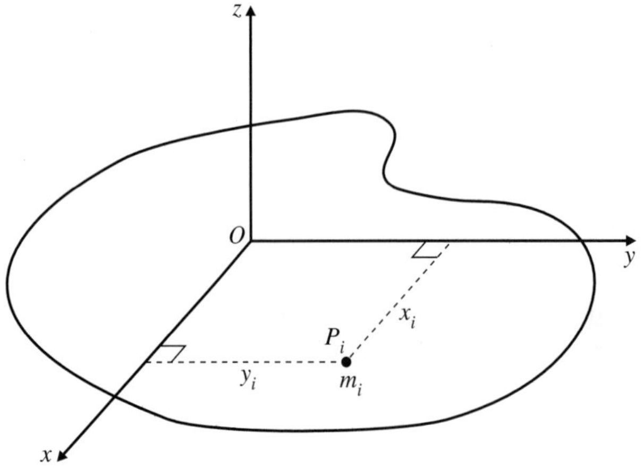

Suppose the moments of inertia of the lamina about the axes Ox and Oy are I_x and I_y respectively.

Then $\quad I_x = \Sigma m_i y_i^2$

and $\quad I_y = \Sigma m_i x_i^2$

So $\quad I_x + I_y = \Sigma m_i y_i^2 + \Sigma m_i x_i^2$

$$= \Sigma m_i (x_i^2 + y_i^2)$$

but $(x_i^2 + y_i^2) = (\text{distance of } P_i \text{ from } O)^2$

$= (\text{distance of } P_i \text{ from } Oz)^2$

Hence $I_x + I_y = \Sigma m_i\ (\text{distance of } P_i \text{ from } Oz)^2$

that is, $I_x + I_y = I_z$

This is the **perpendicular axes theorem for a lamina**:

- **Where I_x, I_y are the moments of inertia of the lamina about two perpendicular axes Ox and Oy in the plane of the lamina and I_z is the moment of inertia of the lamina about the mutually perpendicular axis Oz,**

$$I_x + I_y = I_z$$

Remember that the perpendicular axes theorem applies to *laminae* only.

Example 11

Find the moment of inertia of a uniform rectangular lamina of mass m and sides $2a$, $2b$ about

(a) an axis through its centre G, perpendicular to its plane,
(b) an axis through a corner, perpendicular to its plane.

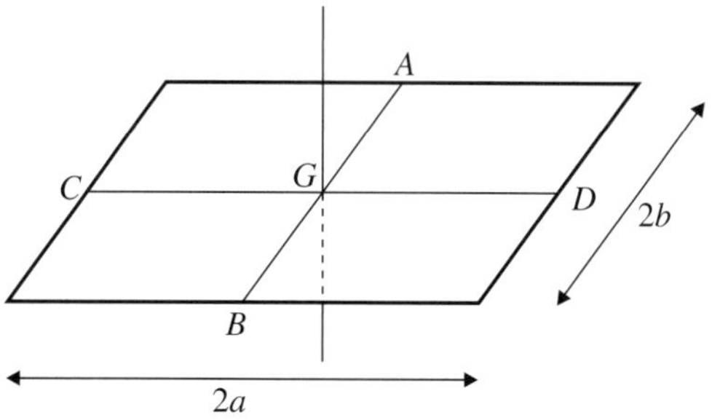

(a) By the stretching rule (section 3.5) the moment of inertia of a rectangular lamina about an axis joining the mid-points of opposite sides is the same as the moment of inertia of a rod of the same mass and length, along an axis through its centre perpendicular to its length.

So M.I. of rectangle about $AB = \frac{1}{3}ma^2$

and M.I. of rectangle about $CD = \frac{1}{3}mb^2$

Hence by the perpendicular axes theorem, M.I. of rectangle about the perpendicular axis through its centre, G, is

$$\tfrac{1}{3}m(a^2 + b^2)$$

(b) An axis through a corner perpendicular to the plane of the lamina is parallel to the axis through its centre G perpendicular to its plane. (This axis passes through the centre of mass of the lamina.) The distance, d, from the centre to a corner is $\sqrt{(a^2+b^2)}$. So, by the parallel axis theorem,

M.I. about perpendicular axis through a corner

$$\begin{aligned}&= \text{M.I. about the parallel axis through } G\text{, the}\\&\quad\ \text{centre of mass} + md^2\\&= \tfrac{1}{3}m(a^2+b^2) + m(a^2+b^2)\\&= \tfrac{4}{3}m(a^2+b^2)\end{aligned}$$

Example 12

A uniform cuboid of mass m has edges of length l, b and h. Calculate the moment of inertia of the cuboid about an edge of length h.

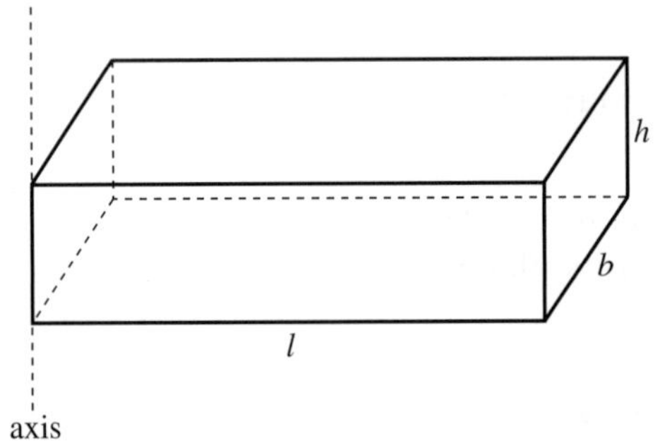

The cuboid can be obtained by stretching a rectangular lamina of sides l, b along an axis through a corner, perpendicular to its plane. The moment of inertia of such a lamina about this axis is, by the result of Example 11:

$$\tfrac{4}{3}m\left[\left(\frac{l}{2}\right)^2 + \left(\frac{b}{2}\right)^2\right] = \tfrac{1}{3}m(l^2+b^2)$$

So the moment of inertia of the cuboid about the side of length h is $\tfrac{1}{3}m(l^2+b^2)$.

The following example uses integration, the parallel axes rule and the perpendicular axes rule.

Example 13

Find the moment of inertia of a uniform right solid cone of mass M, height h and base radius a about

(a) the axis of the cone,
(b) a diameter of the base.

If these two are equal, show that $2h^2 = 3a^2$. [E]

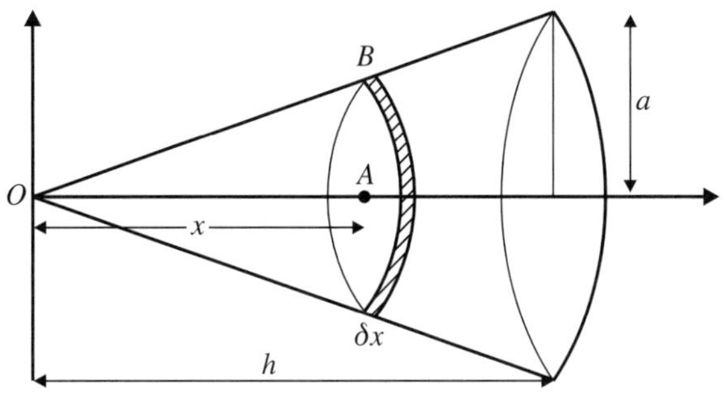

(a) Consider the cone to be composed of discs whose planes are perpendicular to the axis of the cone. Let a typical disc have thickness δx, be distance x from the vertex O of the cone and have centre A, as shown shaded in the diagram.

By similar triangles, the radius AB of the disc is $\dfrac{xa}{h}$.

So $$\text{volume of disc} = \pi\left(\frac{xa}{h}\right)^2 \delta x$$

The mass per unit volume of the cone is $M \div (\frac{1}{3}\pi a^2 h) = \dfrac{3M}{\pi a^2 h}$.

So $$\text{mass of the disc} = \pi\left(\frac{xa}{h}\right)^2 \delta x \times \frac{3M}{\pi a^2 h} = \frac{3Mx^2\delta x}{h^3}$$

The axis passes through the centre of the disc and is perpendicular to the disc.

So $$\begin{aligned}\text{M.I. of disc} &= \tfrac{1}{2} \times \text{mass} \times \text{radius}^2\\ &= \tfrac{1}{2} \times \frac{3Mx^2\delta x}{h^3} \times \left(\frac{xa}{h}\right)^2\\ &= \frac{3Mx^4a^2\delta x}{2h^5}\end{aligned}$$

Adding the moments of inertia of the separate discs and letting $\delta x \to 0$ gives

$$\begin{aligned}\text{M.I. of cone about its axis} &= \int_0^h \frac{3Mx^4a^2}{2h^5}\,\mathrm{d}x\\ &= \frac{3Ma^2}{2h^5}\int_0^h x^4\,\mathrm{d}x\\ &= \frac{3Ma^2}{2h^5}\left[\frac{x^5}{5}\right]_0^h\\ &= \frac{3Ma^2}{2h^5} \times \frac{h^5}{5}\\ &= \frac{3Ma^2}{10}\end{aligned}$$

(b) Remember that the perpendicular axis rule applies *only to laminae*. Hence, although it is used in this part of the question, it cannot be applied to the result of part (a) as a cone is a *solid*. Instead, the moment of inertia of the cone about a diameter of its base must be calculated from first principles.

Consider the cone to be composed of a set of discs as in part (a).

From part (a), the moment of inertia of a typical disc about a perpendicular axis through its centre is $\dfrac{3Mx^4a^2\delta x}{2h^5}$.

So, by the perpendicular axes rule:

$$\text{M.I. of disc about a diameter} = \frac{3Mx^4a^2\delta x}{4h^5}$$

The diameter of the disc is at a distance $(h - x)$ from the diameter of the base of the cone. So, by the parallel axes rule:

M.I. of disc about the diameter of the base of the cone

$$= \frac{3Mx^4a^2\delta x}{4h^5} + \frac{3Mx^2\delta x}{h^3} \times (h - x)^2$$

Adding the moments of inertia of the separate discs and letting $\delta x \to 0$ gives

$$\begin{aligned}
\text{M.I. of cone about diameter of its base} &= \int_0^h \frac{3Mx^4a^2}{4h^5}\,\mathrm{d}x + \int_0^h \frac{3Mx^2(h-x)^2}{h^3}\,\mathrm{d}x \\
&= \frac{3Ma^2}{20} + \frac{3M}{h^3}\int_0^h (h^2x^2 - 2hx^3 + x^4)\,\mathrm{d}x \\
&= \frac{3Ma^2}{20} + \frac{3M}{h^3}\left[\frac{h^2x^3}{3} - \frac{2hx^4}{4} + \frac{x^5}{5}\right]_0^h \\
&= \frac{3Ma^2}{20} + \frac{3M}{h^3}\left[\frac{h^5}{3} - \frac{2h^5}{4} + \frac{h^5}{5}\right] \\
&= \frac{3Ma^2}{20} + \frac{3M}{h^3} \times \frac{h^5}{30} \\
&= \frac{3Ma^2}{20} + \frac{Mh^2}{10}
\end{aligned}$$

If the two moments of inertia are equal, then

$$\begin{aligned}
\frac{3Ma^2}{10} &= \frac{3Ma^2}{20} + \frac{Mh^2}{10} \\
3a^2 &= \frac{3a^2}{2} + h^2 \\
\frac{3a^2}{2} &= h^2 \\
2h^2 &= 3a^2
\end{aligned}$$

Exercise 3B

1 Find, by integration, the moment of inertia of a uniform solid cone of base radius r, height h and mass m about its axis of symmetry.

2 Find the moment of inertia of a uniform circular disc of mass m and radius r about an axis through a point on the rim perpendicular to the plane of the disc.

3 Find the moment of inertia of a uniform disc of mass m and radius r about a diameter. Hence find the moment of inertia of the disc about a tangent.

4 Find the moment of inertia of a uniform rod of length $2a$ and mass m about an axis parallel to the rod and at a distance d from the rod.

5 A uniform square framework is formed by joining four rods of length $2a$ and mass m. Find the moment of inertia of the framework

(a) about an axis through the centre perpendicular to the plane of the framework,

(b) about an axis through one corner perpendicular to the plane of the framework,

(c) about an axis through the mid-point of a side perpendicular to the plane of the framework.

6 A uniform rod of length $2r$ and mass $2m$ has two uniform circular discs of mass m and radius r attached one at each end, so that the rod lies in the plane of the discs as shown.

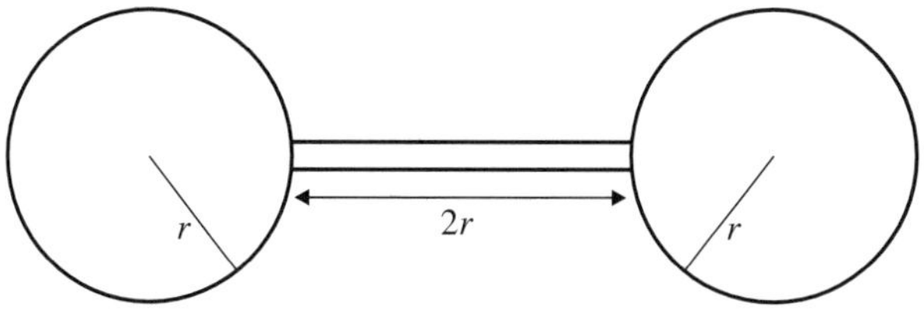

The rod lies along the line of centres of the two discs. Find the moment of inertia of the combined body

(a) about an axis through the mid-point of the rod perpendicular to the plane of the body,

(b) about an axis through one end of the rod perpendicular to the plane of the body,

(c) about an axis through one end of the rod perpendicular to the rod and in the plane of the body,
(d) about an axis through a point of trisection of the rod, perpendicular to the rod and in the plane of the body.

7 Find the moment of inertia of a uniform brick of mass m and sides a, b, c about an axis joining the mid-points of two edges of length a of one face of the brick with edges a and c.

8 Find the radius of gyration of a uniform circular disc radius r about an axis through a point on its rim perpendicular to the disc.

9 Show that if a body has a radius of gyration k_G about an axis l through its centre of mass G, then its radius of gyration about a parallel axis at a distance d from l is $\sqrt{(k_G^2 + d^2)}$.

10 A uniform solid is formed by rotating the region bounded by the curve $y^2 = 16x$ and the line $x = 4$ through one revolution about the x-axis. Find the radius of gyration of this solid about the x-axis.

11 Find the moment of inertia of a uniform solid sphere of mass m and radius r about a tangent to any point on its surface.

12 A uniform solid is formed by rotating the region bounded by the curve with equation $y = \sin x$ and the x-axis between $x = 0$ and $x = \pi$ through one revolution about the x-axis. Find the radius of gyration of this solid about the x-axis.

13 A fly-wheel of mass m is made from a uniform circular disc of radius $2a$ by cutting away a concentric circular disc of radius a. Find the moment of inertia of the fly-wheel
(a) about an axis through the centre perpendicular to the fly-wheel,
(b) about an axis perpendicular to the fly-wheel through a point on its outer rim.

14 Find the radius of gyration of a uniform square lamina of mass m and side $2a$ about a diagonal.

15 A uniform wire of length $12l$ and mass M is bent to form an equilateral triangle. Find the moment of inertia of the triangle about an axis through a vertex perpendicular to the plane of the triangle.

16 A body is formed from a circular hoop of radius r and mass $4m$ and three uniform rods each of length $2r$ and mass m, which are fixed inside the hoop as shown.

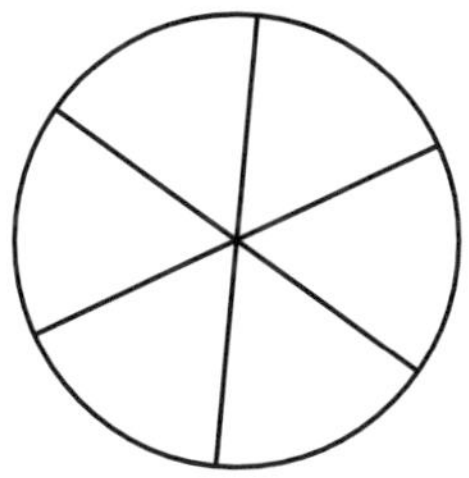

The angles between the rods are all equal. Find the moment of inertia of the body about an axis along one of the rods.

17 An ear-ring of mass m is formed by cutting out a circle of radius r from a thin uniform circular disc of metal of radius $2r$. The centre C of the larger disc lies on the circumference of the hole as shown in the diagram.

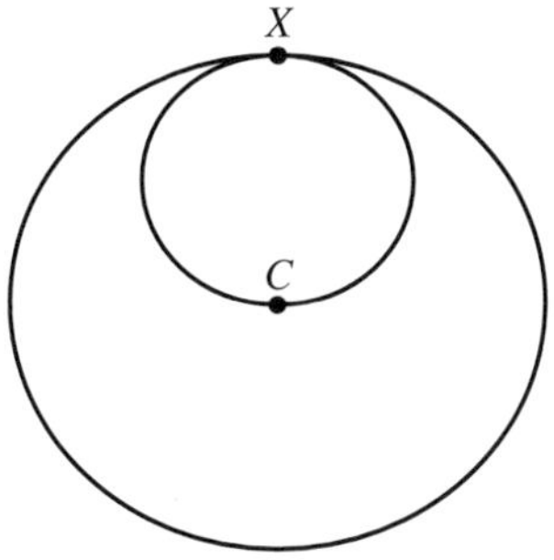

CX is a diameter of the smaller circle. Find the moment of inertia of the ear-ring about an axis through X perpendicular to the plane of the ear-ring.

18 Prove, using integration, that the moment of inertia of a uniform solid sphere, of mass M and radius r, about a diameter is $\frac{2}{5}Mr^2$. Hence obtain the moment of inertia of a uniform solid hemisphere, of mass m and radius r, about a diameter of its plane face.

19 A solid spindle is made by joining the plane faces of two uniform solid cones, both having base radius r. One cone has height h and mass m; the second cone has height $3h$ and mass $5m$. Find the moment of inertia of the spindle about an axis along a diameter of the common plane face of the cones. (You may assume that the moment of inertia of a uniform right solid cone of mass m, base radius a and height h about the axis of the cone is $\frac{3}{10}ma^2$.)

20 (a) By considering a circle to be composed of two semicircles, show that the moment of inertia of a uniform semicircular lamina of mass m and radius r about an axis along its straight edge is $\frac{1}{4}mr^2$.

(b) A sign outside a shop consists of a rectangle of sides a and $2a$ with a semicircle of diameter $2a$ attached to one side as shown in the diagram.

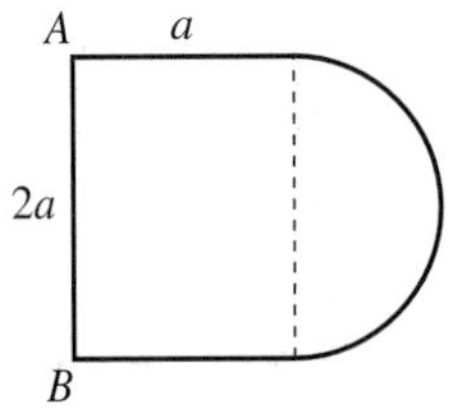

The sign is cut from a thin sheet of plywood and has mass M. Find the moment of inertia of the sign about the side AB.

SUMMARY OF KEY POINTS

1 The moment of inertia (M.I.) of a rigid body about a given axis is $\Sigma m_i r_i^2$ where m_i is the mass of a typical particle and r_i is the distance of that particle from the axis. Moments of inertia are usually measured in kg m^2.

2 **The additive rule**
If two bodies have moments of inertia I_1 and I_2 about the same axis, then the moment of inertia of the composite body about the same axis is $I_1 + I_2$.

3 **The stretching rule**
If one body can be obtained from another body of equal mass by stretching parallel to the axis without alteration of the distribution of mass relative to the axis, then the moments of inertia of the two bodies are the same.

4 The radius of gyration k of a body of mass m about a specified axis is the distance from the axis at which a particle of equal mass must be placed to have the same moment of inertia I about the axis.

$$I = mk^2$$

5 **The parallel axis theorem**
If the moment of inertia of a body of mass m about an axis through its centre of mass, G, is I_G then the moment of inertia about any axis parallel to the original axis and distance d from it is $I_G + md^2$.

6 **The perpendicular axis theorem**
When the moments of inertia of a *lamina* about two perpendicular axes in the plane of the lamina are known, then the moment of inertia of the lamina about an axis through their point of intersection perpendicular to the plane of the lamina is the sum of these two moments of inertia.

7 You may quote the following standard results without proof unless you are asked to prove them:
For uniform bodies of mass m:
Thin rod, length $2l$, about perpendicular axis through centre: $\frac{1}{3}ml^2$
Rectangular lamina about axis in plane bisecting edges of length $2l$: $\frac{1}{3}ml^2$
Thin rod, length $2l$, about perpendicular axis through end: $\frac{4}{3}ml^2$
Rectangular lamina about edge perpendicular to edges of length $2l$: $\frac{4}{3}ml^2$
Rectangular lamina, sides $2a$ and $2b$, about perpendicular axis through centre: $\frac{1}{3}m(a^2 + b^2)$
Hoop or cylindrical shell of radius r about axis through centre: mr^2
Hoop of radius r about a diameter: $\frac{1}{2}mr^2$
Disc or solid cylinder of radius r about axis through centre: $\frac{1}{2}mr^2$
Disc of radius r about a diameter: $\frac{1}{4}mr^2$
Solid sphere, radius r, about diameter: $\frac{2}{5}mr^2$
Spherical shell of radius r about a diameter: $\frac{2}{3}mr^2$

Rotation of a rigid body about a fixed smooth axis

4

If you want to study the motion of a rigid body rotating about a fixed axis, you need to know the moment of inertia of the body about that axis. The methods for calculating moments of inertia have been demonstrated in chapter 3. Now you can study the rotational motion of any rigid body for which you can calculate the moment of inertia about a specified axis. In particular, the list of standard results in the summary of chapter 3 gives the moments of inertia of various uniform bodies about specified axes. So you can study the rotational motion of bodies that can be modelled as some combination of these solids without any complicated initial calculations of their moments of inertia. Pulley wheels, either solid or in the shape of a hoop with spokes, can be modelled as discs if solid or a combination of rods and a hoop otherwise. A solid cylinder can be obtained from a uniform disc by stretching and a cylindrical shell from a circular hoop, combined with uniform discs if the cylinder is closed. By these means, many rotating bodies can be modelled mathematically.

4.1 The kinetic and potential energies of a rotating body

In chapter 3 the kinetic energy of a rigid body rotating with angular speed ω was found to be $\frac{1}{2}\sum_i m_i r_i^2 \omega^2$. When the moment of inertia, I, of the body is defined to be $\sum_i m_i r_i^2$, this gives

■ $$\textbf{K.E. of body} = \tfrac{1}{2} I\omega^2$$

A rotating body will in general have potential energy (P.E.) as well as kinetic energy. Consider the body to be composed of particles of mass $m_1, m_2, \ldots, m_n$ which are at heights $h_1, h_2, \ldots, h_n$ above some (arbitrary) fixed level.

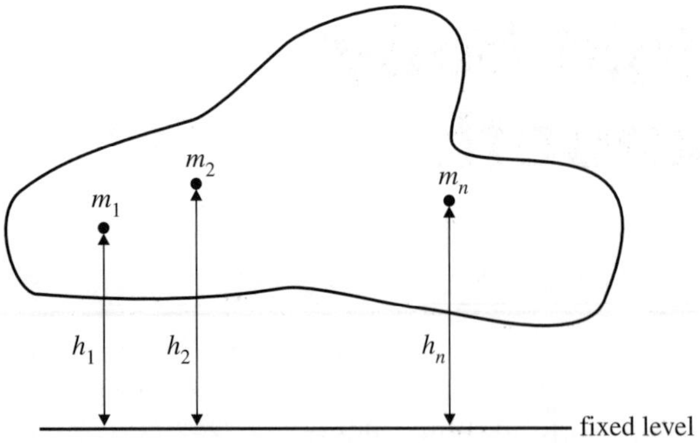

Then, relative to this fixed level, the potential energy of the body is

$$m_1gh_1 + m_2gh_2 + \ldots + m_ngh_n = (m_1h_1 + m_2h_2 + \ldots + m_nh_n)g$$

But

$$m_1h_1 + m_2h_2 + \ldots + m_nh_n = Mh \qquad \text{(Book M2, chapter 2)}$$

where M is the mass of the body and h is the height of the centre of mass of the body above the same fixed level.

So $\qquad$ P.E. of body $= Mgh$

- **When a rigid body is moving, the gain in P.E. of the body is the product of its weight and the vertical height gained by its centre of mass.**

If a rigid body is rotating about a smooth axis and no external forces are acting on the body, the principle of conservation of mechanical energy will hold. That is, the sum of the kinetic and potential energies of the body will remain constant throughout the motion.

Example 1

A uniform circular disc has mass 1 kg and radius 0.5 m. Particles P_1 and P_2 of mass 0.2 kg and 0.5 kg respectively are attached to the disc at distances 0.1 m and 0.3 m respectively from the centre O of the disc. The disc is rotating in a horizontal plane about a smooth vertical axis through its centre O. Calculate the kinetic energy of the system when the disc is rotating at 5 rad s^{-1}.

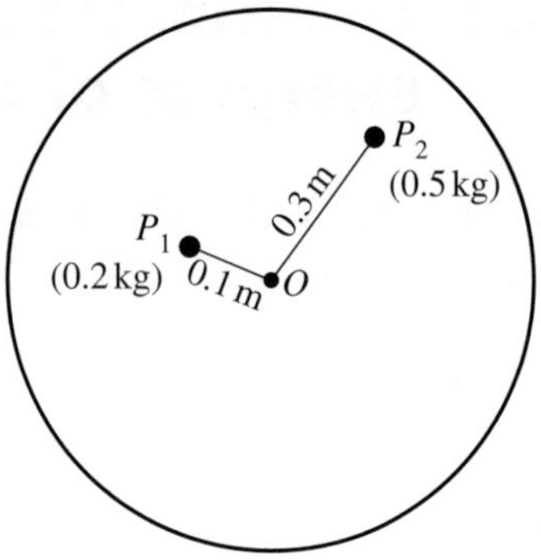

By the additive rule:

$$\text{M.I. of system, } I = (\tfrac{1}{2} \times 1 \times 0.5^2 + 0.2 \times 0.1^2 + 0.5 \times 0.3^2)\,\text{kg m}^2$$
$$= 0.172\,\text{kg m}^2$$

$$\text{K.E. of system} = \tfrac{1}{2}I\omega^2 = \tfrac{1}{2} \times 0.172 \times 5^2\,\text{J}$$
$$= 2.15\,\text{J}$$

The kinetic energy of the disc and particles is 2.15 J.

Example 2

A uniform rod AB of length 1 m and mass 0.5 kg is free to rotate in a vertical plane about a fixed, smooth, horizontal axis through A. It is released from rest with AB horizontal.

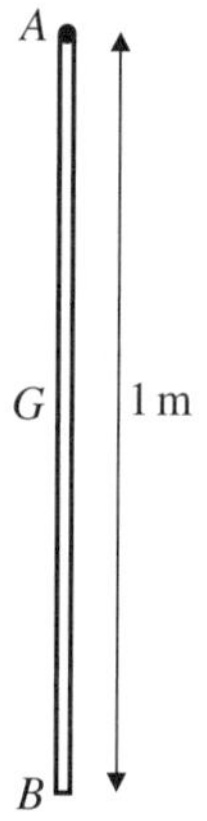

(a) Calculate the potential energy lost by the rod in falling to the position where it is vertical with B below A.
(b) Calculate the angular speed of the rod at this instant.

(a) The centre of mass G of the rod is at its mid-point.

So: P.E. lost = mass $\times\, g \,\times$ distance fallen by centre of mass

$$= 0.5 \times 9.8 \times 0.5\,\text{J}$$
$$= 2.45\,\text{J}$$

The potential energy lost is 2.45 J.

(b) Let the angular speed of the rod when AB is vertical be $\omega\,\text{rad s}^{-1}$.

Use the standard result, $\frac{1}{3}ma^2$, for the moment of inertia of a uniform rod with mass m and length $2a$ about an axis through its mid-point, perpendicular to its length:

M.I. of rod about axis through its centre of mass G

$$= \tfrac{1}{3} \times 0.5 \times 0.5^2\,\text{kg m}^2$$

Distance from G to end of rod = 0.5 m.

So, by the parallel axis rule:

M.I. of rod about given axis through A

$$= (\tfrac{1}{3} \times 0.5^3 + 0.5 \times 0.5^2)\,\text{kg m}^2$$
$$= \tfrac{4}{3} \times 0.5^3\,\text{kg m}^2$$
$$= \tfrac{4}{3} \times \tfrac{1}{2} \times \tfrac{1}{2} \times \tfrac{1}{2}\,\text{kg m}^2$$
$$= \tfrac{1}{6}\,\text{kg m}^2$$

So K.E. of rod $= \frac{1}{2}I\omega^2 = \frac{1}{2} \times \frac{1}{6}\omega^2\,\text{J}$

By the principle of conservation of mechanical energy:

K.E. gained by rod = P.E. lost by rod

$$\tfrac{1}{12}\omega^2 = 2.45$$
$$\omega^2 = 12 \times 2.45$$
$$\omega = 5.42$$

The angular speed of the rod is $5.42\,\text{rad s}^{-1}$.

Example 3

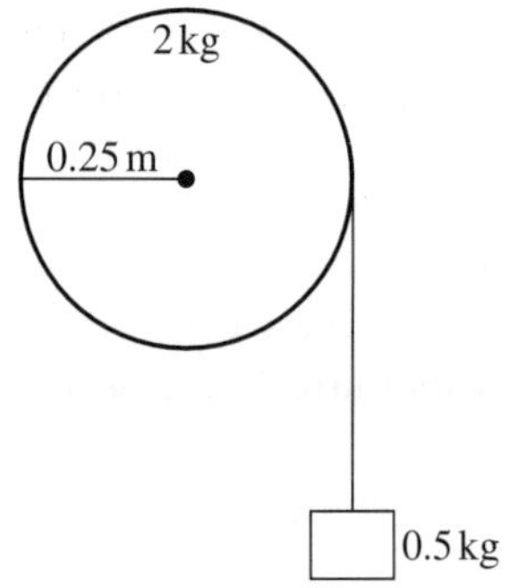

A pulley wheel has mass 2 kg and radius 0.25 m. One end of a rope is attached to a point on the rim of the wheel and the rope is wound several times around the rim. A brick of mass 0.5 kg attached to the other end of the rope hangs freely. The brick is released from rest. Assuming the axis of rotation of the wheel is horizontal, perpendicular to the wheel, and passes through the centre of the wheel, determine how far the brick must descend before it has acquired a speed of $2\,\text{m}\,\text{s}^{-1}$. State clearly any additional assumptions you make in order to model this situation.

Assume that the axis is smooth and so there are no external forces acting.

Assume that the pulley wheel can be modelled as a uniform disc and the brick as a particle.

Assume that the rope is light and inextensible.

When the brick has a speed of $2\,\text{m}\,\text{s}^{-1}$, the (linear) speed of a point on the rim of the pulley wheel must be $2\,\text{m}\,\text{s}^{-1}$.

The angular speed ω and the linear speed v are related by the equation

$$v = r\omega$$

where r is the radius of the disc.

So $$2 = 0.25\omega$$

Hence the angular speed is $8\,\text{rad}\,\text{s}^{-1}$.

$$\begin{aligned}\text{M.I. of wheel} &= \tfrac{1}{2}mr^2\\ &= \tfrac{1}{2}\times 2\times 0.25^2\,\text{kg}\,\text{m}^2\\ &= 0.25^2\,\text{kg}\,\text{m}^2\end{aligned}$$

So $$\begin{aligned}\text{K.E. gained by wheel} &= \tfrac{1}{2}I\omega^2\\ &= \tfrac{1}{2}\times 0.25^2\times 8^2\,\text{J}\\ &= 2\,\text{J}\end{aligned}$$

$$\begin{aligned}\text{K.E. gained by brick} &= \tfrac{1}{2}mv^2\\ &= \tfrac{1}{2}\times 0.5\times 2^2\,\text{J}\\ &= 1\,\text{J}\end{aligned}$$

$$\text{P.E. lost by brick} = mgh = 0.5\times 9.8h\,\text{J}$$

where h m is the distance the brick has fallen.

By the conservation of mechanical energy:

$$\text{P.E. lost} = \text{K.E. gained}$$

So $$0.5 \times 9.8h = 2 + 1$$

$$h = \frac{3}{0.5 \times 9.8} = 0.612$$

The brick must descend a distance 0.612 m.

Exercise 4A

1 A uniform circular hoop of mass 2 kg and radius 0.5 m is rotating in a horizontal plane about a smooth vertical axis through a point on its circumference. Calculate its kinetic energy when it is rotating at 4 rad s^{-1}.

2 A uniform circular disc of mass 3 kg and radius 0.4 m has particles of mass 0.1 kg, 0.4 kg and 1 kg attached to points at distances 0.2 m, 0.2 m and 0.3 m respectively from the centre of the disc. The disc is rotating in a horizontal plane about a smooth vertical axis through its centre. Calculate the kinetic energy of the loaded disc when it is rotating at 3 rad s^{-1}.
The disc is now brought to rest. Calculate the work done by the retarding force.

3 A uniform rod of length 1.2 m and mass 0.8 kg has particles of mass 0.2 kg and 0.5 kg attached, one at each end. The rod is rotating in a horizontal plane about a smooth vertical axis. Given that the rod is rotating with angular speed 5 rad s^{-1}, calculate the kinetic energy of the rod
(a) when the axis passes through the mid-point of the rod,
(b) when the axis passes through the centre of mass of the loaded rod.

4 A uniform rod AB of length 1.5 m and mass 2 kg is smoothly pivoted at A. It is initially at rest with B vertically above A. It is slightly disturbed and rotates in a vertical plane.
(a) Calculate the potential energy lost by the rod when it reaches the horizontal position.
(b) Calculate the angular speed of the rod as it passes through the horizontal position.
(c) Calculate the angular speed of the rod when B is vertically below A.

5 A pulley wheel of mass 5 kg and radius 2 m has one end of a rope attached to a point of its rim and wound several times around its rim. A brick of mass 2 kg attached to the other end of the rope hangs freely 3 m above the ground. The brick is released from rest. Assuming that the pulley wheel can be modelled as a uniform disc which can rotate in a vertical plane about a smooth horizontal axis through its centre, and the rope can be modelled as a light inextensible string, calculate the angular speed of the wheel when the brick hits the ground. State the model used for the brick in this calculation.

6 A uniform rod AB of mass m and length $4l$ is free to rotate in a vertical plane about a smooth horizontal axis through its mid-point. Particles of mass $3m$ and m are attached to ends A and B respectively.

(a) The rod is held with AB horizontal and then released from rest. Find, in terms of l and g, the angular speed of the rod when AB is vertical.

(b) If *instead* the rod is initially vertical with A below B and is then given an angular speed of $\sqrt{\left(\frac{g}{2l}\right)}$, calculate the angle between AB and the downward vertical when the rod first comes to rest.

7 A uniform disc of mass 1.5 kg and radius 1.5 m is free to rotate in a vertical plane about a smooth horizontal axis through its centre. A rope is attached to a point on the rim of the disc and wound around the disc. The disc is initially at rest. The free end of the rope is then pulled in a direction in the plane of the disc as shown in the diagram. Given that the tension in the rope is 10 N, calculate the angle the disc has turned through when the angular speed is 3 rad s^{-1}.

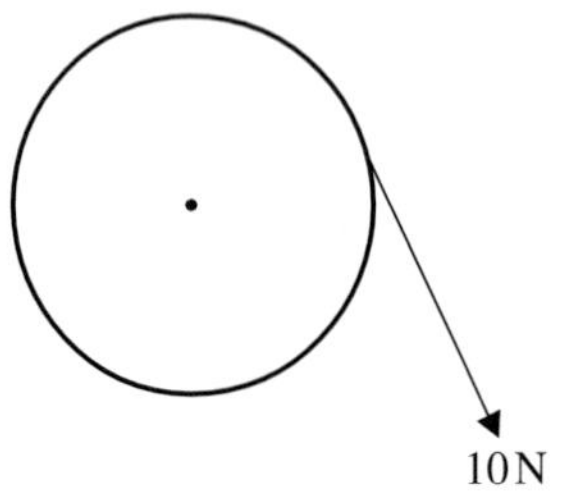

8 A uniform rod AB of mass m and length $2l$ has a particle of mass $2m$ attached to end B. The rod is free to rotate in a vertical plane about a smooth horizontal axis through the point C of the rod where $AC = \frac{l}{3}$. When the rod is hanging in equilibrium with B vertically below A it is given an angular speed of ω.

(a) Find the least possible value of ω if the rod is to perform complete revolutions.

(b) If ω is twice this minimum value, find the speed of the particle as it passes through its highest point.

(c) If ω is half this minimum value find the angle between BC and the downward vertical when the particle is at its highest point.

9 A uniform rod AB of mass m has a particle of mass m attached at B. The rod rotates in a vertical plane about a smooth horizontal axis through A. The greatest angular speed of the rod is $2\sqrt{\left(\frac{g}{l}\right)}$.

(a) Given that the rod just makes complete revolutions, find the length of the rod.

(b) As end B passes through its lowest point a variable force is applied to the rod. The rod continues to rotate and B passes vertically above A with an angular speed of $\frac{3}{2}\sqrt{\left(\frac{g}{l}\right)}$. Calculate the work done by the force in moving B from the lowest point to the highest point of its path.

10 A uniform wire in the form of a circle of radius a is free to rotate about a smooth horizontal axis through a point A of its circumference and perpendicular to its plane. It is released from rest with the diameter AB horizontal. Find its angular speed when AB is vertical.

11 A bucket of mass m is attached to the end of a light rope, the other end of which is attached to the circumference of a wheel of mass M. The rope is wound several times around the wheel. The wheel can rotate freely about a smooth fixed horizontal axis through its centre. Assuming that the entire mass of the wheel is concentrated in its rim, calculate the speed of the bucket when it has fallen a distance h from rest.

12 The figure shows an advertising sign in the form of a rectangular plate of sides $2a$ and $2b$. It is free to rotate about a smooth fixed horizontal axis which coincides with the side AB.

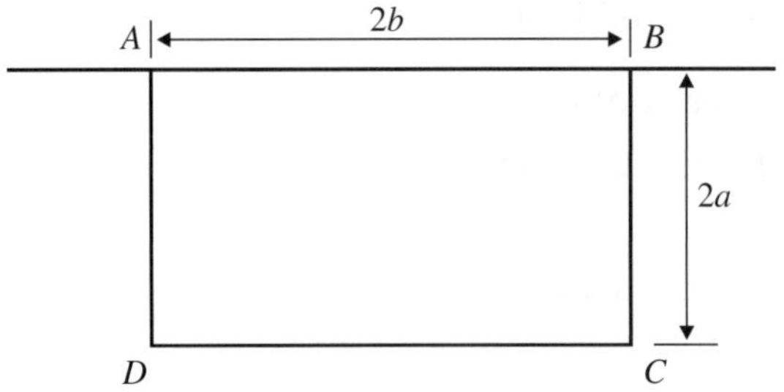

When the sign is hanging freely below the axis it is given an angular speed of $\left(\frac{kg}{a}\right)^{\frac{1}{2}}$. Determine the range of values of k for which the sign makes complete revolutions. State the mathematical model used for the sign in the calculation.

13 A ring of mass m and radius a has a particle of mass m attached to it at the point A. The ring can rotate about a fixed smooth horizontal axis which is tangential to the ring at the point B which is diametrically opposite A. The system is released from rest with AB horizontal. Find the angular speed of the system when AB has turned through an angle $\frac{\pi}{6}$.

14

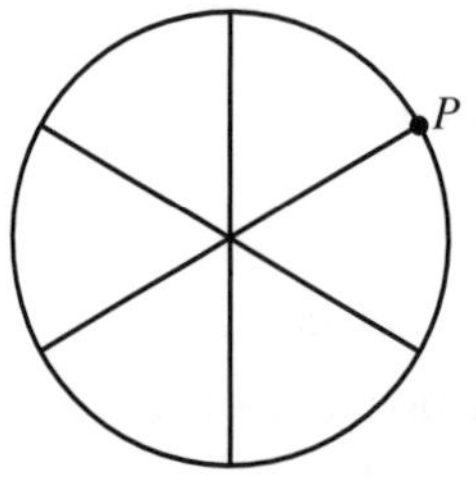

A flywheel is made from a circular hoop of radius r and mass $4M$ together with three equally spaced rods, each of length $2r$ and mass M. A particle P of mass $3M$ is attached to the circumference of the flywheel, at the end of one rod. The flywheel is free to rotate about a smooth horizontal axis through its centre perpendicular to its plane. The flywheel is released from rest with one rod vertical and P above the horizontal level of the axis, as shown in the diagram. Find, in terms of r and g, the angular speed of the flywheel when P is vertically below the axis.

15 A uniform circular disc has mass $3m$ and radius r. A particle of mass m is attached to the disc at point A of its circumference. The loaded disc is free to rotate about a horizontal axis which is tangential to the disc at the point B, where AB is a diameter. The disc is released from rest with AB at an angle of 60° with the upward vertical. Find the angular speed of the disc when AB is vertical.

16 A solid cone of mass m has base radius r and height h. It is free to rotate about a fixed horizontal axis which coincides with a diameter of its base. It is initially at rest with its vertex vertically above the axis. The cone is slightly disturbed and begins to rotate. Find the angular speed of the cone when its vertex is vertically below the axis.

4.2 The equation of rotational motion

Consider a rigid body which is rotating about a fixed smooth axis which passes through the body. Think of the body as composed of particles $P_1, P_2, \ldots, P_n$, of masses $m_1, m_2, \ldots, m_n$, which are at distances $r_1, r_2, \ldots, r_n$ from the axis. The diagram shows a plane section of the body, perpendicular to the axis and containing the particle P_i. The axis of rotation passes through the point O of this section. When the body is rotating with angular speed $\dot{\theta}$ each line OP_i is also rotating with angular speed $\dot{\theta}$. The particle shown is therefore moving in a circle, centre on the axis, radius r_i, with angular speed $\dot{\theta}$. As was shown in Book M3, chapter 4, the acceleration of P_i has components $r_i\dot{\theta}^2$ along P_iO and $r_i\ddot{\theta}$ perpendicular to OP_i, respectively. Let the force acting on P_i have component N_i in the direction perpendicular to OP_i. Then applying Newton's second law to P_i in this direction gives

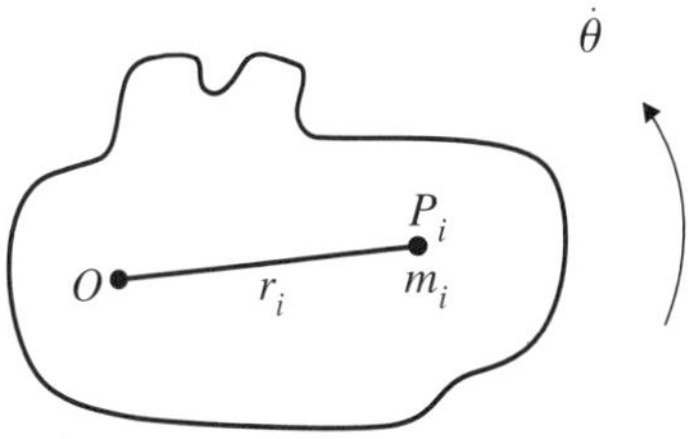

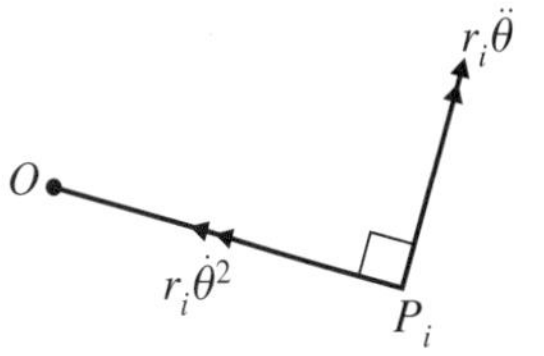

$$N_i = m_i r_i \ddot{\theta}$$

and multiplying by r_i:

$$N_i r_i = m_i r_i^2 \ddot{\theta}$$

As P_i is just one constituent particle of the rigid body, summing over all the particles $P_1, P_2, \ldots, P_n$ gives, for the whole body,

$$\sum_i N_i r_i = \sum_i m_i r_i^2 \ddot{\theta}$$

Since the body is rigid, that is, the particles cannot move relative to one another, $\ddot{\theta}$ is the same for all the particles. So

$$\sum_i N_i r_i = \left(\sum_i m_i r_i^2\right)\ddot{\theta}$$

and $\sum_i m_i r_i^2$ is just the moment of inertia, I, of the body about the axis through O, so

$$\sum_i N_i r_i = I\ddot{\theta}$$

$\sum_i N_i r_i$ is the sum of the moments about the axis of rotation of the forces acting on each particle of the body (Book M1, chapter 6). It is the same as the moment about the axis of rotation of the resultant force acting on the body. This moment can be denoted by L, giving the **equation of rotational motion**:

■ $$\boldsymbol{L = I\ddot{\theta}}$$

where L is the moment of the resultant force on a body about the axis of rotation, I is the moment of inertia of the body about the axis and $\ddot{\theta}$ is the angular acceleration of the body.

Example 4

A pulley wheel has mass 1 kg and radius 0.25 m. A rope has one end attached to a point of the rim of the wheel and is wound several times around the rim. A brick of mass 0.5 kg attached to the other end of the rope hangs freely. The wheel is free to rotate in a vertical plane about a fixed, smooth, horizontal axis through the centre of the wheel. The brick is released from rest. By modelling the pulley wheel as a uniform disc, the brick as a particle and the rope as a light, inextensible string, calculate the tension in the rope and the acceleration of the brick.

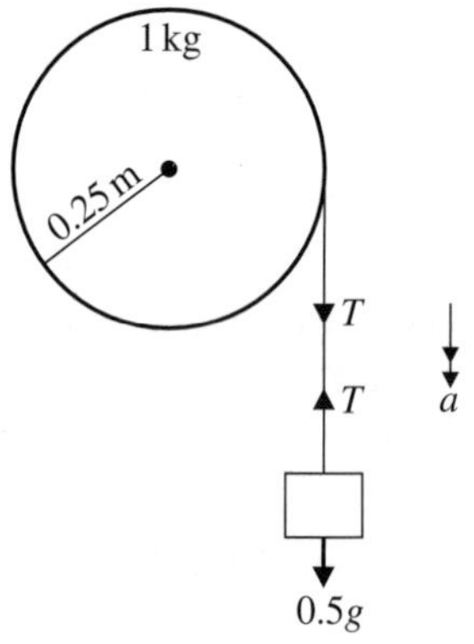

Let the tension be T N and the acceleration of the brick be a m s^{-2}.

As in the motion of two connected particles, you must consider the motions of the disc and the brick separately.

For the brick, using the equation of motion, $F = ma$, gives

$$0.5g - T = 0.5a \qquad (1)$$

For the pulley wheel, the linear acceleration of a point on the rim is a m s^{-2}. Hence the angular acceleration, $\ddot{\theta}$ rad s^{-2}, of the wheel is given by

$$r\ddot{\theta} = a$$

and since $r = 0.25\,\text{m}$,

$$\ddot{\theta} = \frac{a}{0.25}$$

The moment of inertia of the wheel is $\frac{1}{2} \times 1 \times 0.25^2\,\text{kg m}^2$. Using the equation of rotational motion, $L = I\ddot{\theta}$, gives

$$T \times 0.25 = \tfrac{1}{2} \times 1 \times 0.25^2 \times \frac{a}{0.25}$$

so $$T = \tfrac{1}{2}a$$

Substituting for a in equation (1) gives

$$0.5g - T = T$$
$$2T = 0.5 \times 9.8$$
$$T = 2.45$$

and so $$a = 2T = 2 \times 2.45 = 4.9$$

The tension is 2.45 N and the acceleration is $4.9\,\text{m s}^{-2}$.

4.3 The force on the axis of rotation

To be able to calculate the force exerted by a rotating body on the axis of rotation we must first study the motion of the centre of mass of a system of particles. Consider a system of moving particles $P_1, P_2, \ldots, P_n$ of masses $m_1, m_2, \ldots, m_n$. Let the resultant force acting on P_i be $\mathbf{F}_i$ and let P_i have position vector $\mathbf{r}_i$ relative to some fixed origin O.

Applying Newton's second law to P_i gives

$$\mathbf{F}_i = m_i\ddot{\mathbf{r}}_i$$

So for the whole system

$$\sum_i \mathbf{F}_i = \sum_i m_i\ddot{\mathbf{r}}_i \qquad (1)$$

Let $\sum_i m_i = M$ and let the centre of mass have position vector $\mathbf{r}_G$ given by

$$\sum_i m_i\mathbf{r}_i = M\mathbf{r}_G \qquad (2)$$

Differentiating equation (2) twice with respect to time gives

$$\sum_i m_i\ddot{\mathbf{r}}_i = M\ddot{\mathbf{r}}_G \qquad (3)$$

as m_i and M are constants.

Eliminating $\sum_i m_i \ddot{\mathbf{r}}_i$ from equations (1) and (3) gives

$$\sum_i \mathbf{F}_i = M\ddot{\mathbf{r}}_G$$

But $\sum_i \mathbf{F}_i$ is the resultant of all the forces acting on the system, so

- **The centre of mass of a system of particles has the same acceleration as a particle of mass equal to the total mass of the system situated at the centre of mass under the action of a force equal to the resultant of all the forces acting on the system.**

You can treat a rigid body as if it were a system of particles and so this principle applies. Hence:

- **For a rigid body which is rotating about a fixed smooth axis, the force exerted by the axis on the body can be calculated by considering the motion of a particle of the same mass as the body placed at the centre of mass of the body under the action of the same forces as those acting on the body.**

Suppose a body of mass m is rotating about a smooth, fixed, horizontal axis and that its centre of mass G is a distance r from the axis of rotation. The diagram shows the vertical plane through the centre of mass with the axis of rotation passing through O.

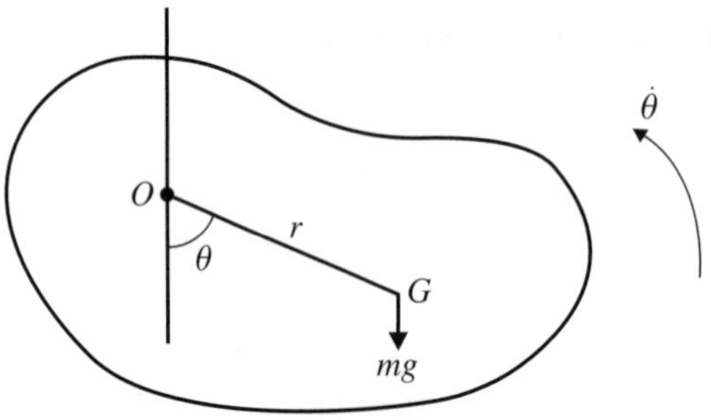

Let the angular speed of the body be $\dot{\theta}$ and consider a particle of mass m placed at G. The particle is moving in a vertical circle, radius r, with angular speed $\dot{\theta}$. The angle between OG and the downward vertical is θ.

The particle's acceleration has components $r\dot{\theta}^2$ and $r\ddot{\theta}$ parallel to GO and perpendicular to OG respectively:

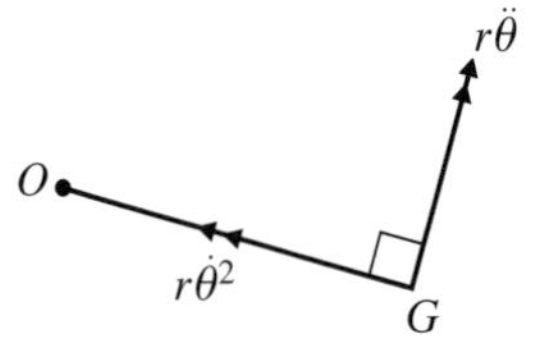

Let the force *from* the axis have components X and Y parallel to GO and perpendicular to OG. The mass of the body is m and the forces acting are X, Y and mg. The equation of motion parallel to GO gives

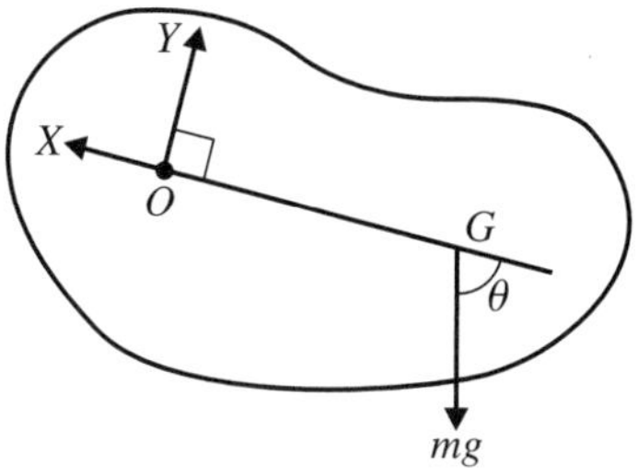

$$X - mg\cos\theta = mr\dot{\theta}^2$$

The equation of motion perpendicular to OG gives

$$Y - mg\sin\theta = mr\ddot{\theta}$$

The value of $r\dot{\theta}^2$ can be found for any position of the body by using the principle of conservation of mechanical energy. The value of $r\ddot{\theta}$ can be found from the equation of rotational motion. Alternatively, differentiating $r\dot{\theta}^2$ with respect to θ gives

$$\frac{\mathrm{d}}{\mathrm{d}\theta}(r\dot{\theta}^2) = 2r\dot{\theta}\frac{\mathrm{d}\dot{\theta}}{\mathrm{d}\theta} \quad \text{by the chain rule (Book P3, chapter 2)}$$
$$= 2r\dot{\theta}\frac{\mathrm{d}\dot{\theta}}{\mathrm{d}t}\cdot\frac{\mathrm{d}t}{\mathrm{d}\theta}$$

and as $\dot{\theta} = \frac{\mathrm{d}\theta}{\mathrm{d}t}$ and $\frac{\mathrm{d}\dot{\theta}}{\mathrm{d}t} = \frac{\mathrm{d}^2\theta}{\mathrm{d}t^2} = \ddot{\theta}$ this gives

$$\frac{\mathrm{d}}{\mathrm{d}\theta}(r\dot{\theta}^2) = 2r\ddot{\theta}$$

Hence the value of $\ddot{\theta}$ can be found by differentiating with respect to θ an expression for $r\dot{\theta}^2$ in terms of θ. Thus you can calculate the components X and Y of the force exerted *by the axis on the body*. Usually you are asked for the force exerted *on the axis*, and so you must use Newton's third law as a final step.

Example 5

A uniform circular disc, centre C, of mass m and radius r can rotate in a vertical plane about a smooth horizontal axis perpendicular to its plane through a point A on its rim. Initially, the disc is held at rest with AC horizontal. It is then released. Find the components of the force on the axis when AC makes an angle θ with the downward vertical.

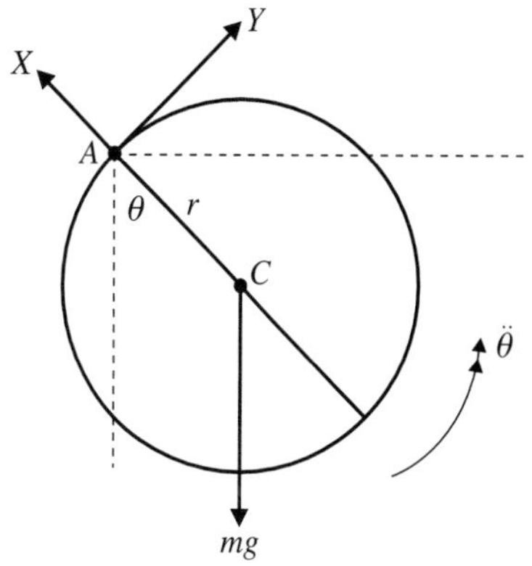

Calculate the magnitude of the force on the axis (a) when AC is vertical, and (b) when $\theta = \frac{\pi}{3}$.

The angular acceleration of the disc is $\ddot{\theta}$.

Let the force from the axis of rotation have components X and Y along and perpendicular to CA, as shown.

Consider the motion of a particle of mass m, placed at the centre of mass C of the disc, when the resultant force acting on the particle is the same as the resultant force acting on the disc. This particle will be moving in a vertical circle, centre A, radius r. It will therefore have accelerations $r\ddot{\theta}$ perpendicular to AC and $r\dot{\theta}^2$ along CA:

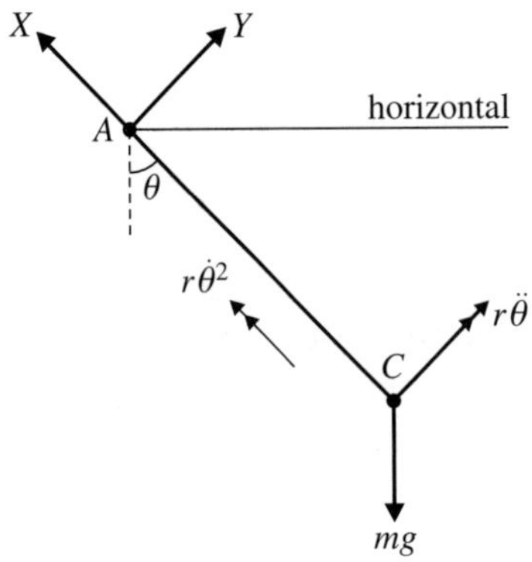

Applying Newton's second law parallel to CA gives

$$X - mg\cos\theta = mr\dot{\theta}^2 \qquad (1)$$

and Newton's second law perpendicular to CA gives

$$Y - mg\sin\theta = mr\ddot{\theta} \qquad (2)$$

You can find $r\dot{\theta}^2$ and $r\ddot{\theta}$ by considering the rotational motion of the disc.

M.I. of disc about axis through centre $C = \frac{1}{2}mr^2$

M.I. of disc about axis through $A = \frac{1}{2}mr^2 + mr^2 = \frac{3}{2}mr^2$

By the principle of conservation of mechanical energy:

K.E. gained by disc = P.E. lost by disc

P.E. lost $= mgr\cos\theta$

K.E. gained $= \frac{1}{2}I\dot{\theta}^2 = \frac{1}{2} \times \frac{3}{2}mr^2\dot{\theta}^2$

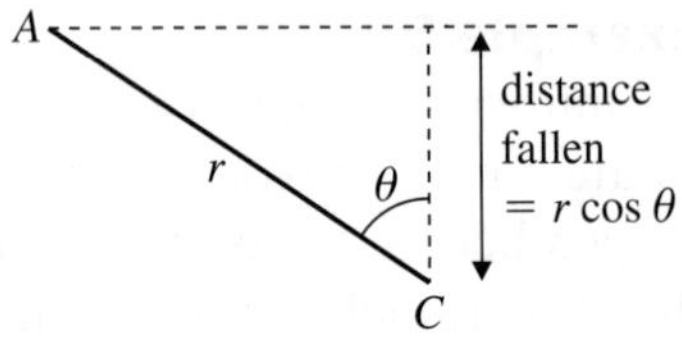

So $$\tfrac{3}{4}mr^2\dot{\theta}^2 = mgr\cos\theta$$

$$r\dot{\theta}^2 = \tfrac{4}{3}g\cos\theta \qquad (3)$$

Applying the equation of rotational motion:

moment of resultant force about axis $= I\ddot{\theta}$

$$mgr\sin\theta = -\tfrac{3}{2}mr^2\ddot{\theta}$$

so $$r\ddot{\theta} = -\tfrac{2}{3}g\sin\theta$$

Alternatively, differentiating equation (3) with respect to θ gives

$$2r\ddot{\theta} = -\tfrac{4}{3}g\sin\theta$$

and so $$r\ddot{\theta} = -\tfrac{2}{3}g\sin\theta$$

Substituting $r\dot{\theta}^2 = \frac{4}{3}g\cos\theta$ and $r\ddot{\theta} = -\frac{2}{3}g\sin\theta$ in equations (1) and (2) gives

$$X - mg\cos\theta = \tfrac{4}{3}mg\cos\theta$$
$$Y - mg\sin\theta = -\tfrac{2}{3}mg\sin\theta$$

and so $\qquad X = \frac{7}{3}mg\cos\theta, \quad Y = \frac{1}{3}mg\sin\theta$

But X and Y are the components of the force exerted *by* the axis *on* the disc. By Newton's third law, the force exerted *on* the axis *by* the disc has components of equal magnitude but opposite direction.

So the force on the axis has components:

$$\tfrac{7}{3}mg\cos\theta \text{ along } AC$$

and

$$\tfrac{1}{3}mg\sin\theta \text{ perpendicular to } AC$$

(in the direction of decreasing θ).

(a) When AC is vertical, $\theta = 0$. So the magnitude of the force on the axis is $\frac{7}{3}mg$. Note that the component of the force perpendicular to AC is zero when AC is vertical.

(b) When $\theta = \frac{\pi}{3}$ the components are

$$\tfrac{7}{3}mg \times \tfrac{1}{2} = \frac{7mg}{6} \text{ along } AC$$

and $\qquad \frac{1}{3}mg \times \frac{\sqrt{3}}{2} = \frac{mg\sqrt{3}}{6}$ perpendicular to AC

So the resultant force has magnitude

$$\sqrt{\left\{\frac{49m^2g^2}{36} + \left(m^2g^2 \times \frac{3}{36}\right)\right\}} = mg\sqrt{\left(\frac{52}{36}\right)} = \frac{mg\sqrt{13}}{3}$$

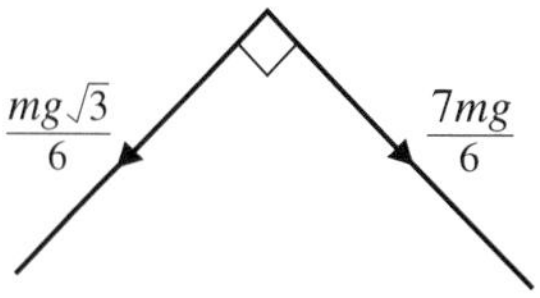

Exercise 4B

1 A uniform rod of length $2a$ and mass m is free to rotate in a horizontal plane about a fixed vertical axis through one end. A horizontal force of constant magnitude $5mg$ is applied to its free end at right angles to the rod. Find the magnitude of the resulting angular acceleration.

2 A uniform circular disc of radius 0.2 m and mass 0.5 kg is free to rotate in a horizontal plane about a smooth, fixed, vertical axis through its centre. A horizontal force of constant magnitude is applied at a point on the rim of the disc in the direction of a tangent to the disc. The disc rotates with angular acceleration 5 rad s^{-2}. Calculate the magnitude of the force.

3 A pulley wheel of mass 2 kg has one end of a rope attached to a point of its rim and the rope is wound several times around the rim. The wheel is free to rotate about a fixed, smooth, horizontal axis through its centre and perpendicular to the wheel. A brick hangs freely, attached to the other end of the rope. The brick is released from rest and falls 10 m in the first two seconds after its release. Assuming that the pulley can be modelled as a uniform circular disc and the brick as a particle, calculate the tension in the rope and the mass of the brick.

4 A uniform rod AB of mass $2m$ and length $3a$ is free to rotate in a vertical plane about a fixed, smooth, horizontal axis through end A of the rod. A particle of mass m is attached to end B. The rod is released from rest with AB horizontal. Find (a) the initial angular acceleration of the system, (b) the angular acceleration when the rod has turned through an angle θ.

5 A uniform solid cylinder of radius 0.5 m and mass 0.5 kg can rotate freely about a smooth, fixed, horizontal axis which coincides with the axis of the cylinder. A light string passes over the cylinder in a vertical plane perpendicular to the axis of rotation. Particles of masses 0.25 kg and 0.5 kg are attached one to each end of the string, and the system is released from rest. Assuming that the string does not slip on the cylinder and that neither particle has reached the cylinder, calculate the tensions in the two parts of the string and the angular acceleration of the cylinder.

6 A uniform rod AB of mass m and length $2a$ is free to rotate in a vertical plane about a fixed, smooth, horizontal axis through the point C of the rod where $AC = \frac{1}{2}a$. The rod is released from rest with AB horizontal. Calculate the magnitude of the force exerted on the axis (a) when AB is vertical with B below A, (b) when AB makes an angle of 60° with the downward vertical.

7 A uniform rod AB of mass m and length $8a$ is free to rotate in a vertical plane about a smooth, fixed, horizontal axis through the point C of the rod where $AC = 2a$. The rod is initially at rest with A vertically below C but is then slightly disturbed and begins to rotate. Find (a) the angular speed when the rod has turned through an angle θ, (b) the magnitude of the force on the axis when the rod is vertical with A vertically above C.

8 A uniform square lamina $ABCD$ of mass m and side $2a$ is free to rotate about a smooth, horizontal axis through A. The axis is perpendicular to the plane of the lamina. The lamina is in equilibrium with C vertically below A when it is given an angular speed $2\sqrt{\left(\frac{g}{a}\right)}$. Show that the lamina will perform complete revolutions and find the horizontal and vertical components of the reaction at A (a) when AC is horizontal, (b) when C is vertically above A.

9 A uniform circular disc, centre O, of mass $2m$ and radius r has a particle of mass m attached to a point A of its surface where $OA = \frac{1}{2}r$. The disc is free to rotate about a fixed, smooth horizontal axis through O, perpendicular to the plane of the disc. The disc is held at rest with A vertically above O. The disc is slightly disturbed from its position of rest. Find the horizontal and vertical components of the force on the axis (a) when OA is horizontal, (b) when A is vertically below O.

10 A uniform rod PQ of mass m and length l is free to rotate in a vertical plane about a fixed smooth horizontal axis passing through P. The rod is released from rest when PQ makes an angle of 60° with the downward vertical. Find
(a) the angular speed of PQ as Q passes through the lowest point of its path,
(b) the magnitude of the force exerted by the rod on the axis at this moment.

11 A uniform rod PQ of length $4l$ and mass m is free to rotate in a vertical plane about a smooth horizontal axis, perpendicular to the rod, through point R of the rod where $PR = l$. A particle of mass $5m$ is attached to the rod at S where $PS = 3l$. The rod is released from rest when PQ is horizontal. Find
(a) the angular speed of the rod at the moment when PQ is vertical,
(b) the magnitude of the force exerted by the rod on the axis when PQ is vertical.

12 A pulley wheel of radius r and mass m is free to rotate about a fixed smooth horizontal axis through its centre perpendicular to the plane of the pulley. A light inextensible string passes over the pulley and carries particles of mass $2m$ and $4m$, one at each end of the string. It can be assumed that the pulley wheel can be modelled as a uniform circular disc and that the string does not slip on the pulley. Find, in terms of r, m, g and π,
(a) the tension in each part of the string,
(b) the magnitude of the acceleration of the particles.

13 A uniform circular disc has mass $4m$ and radius r. A particle of mass m is attached to the disc at point A of its circumference. The loaded disc is free to rotate about a horizontal axis which is tangential to the disc at the point B, where AB is a diameter. The disc is released from rest with AB at an angle of $60°$ with the upward vertical. Find the magnitude of the force exerted by the disc on the axis when AB makes an angle of $60°$ with the downward vertical.

4.4 Angular momentum

As before, consider a rigid body rotating about a fixed axis. Consider the body to be composed of particles $P_1, P_2, \ldots, P_n$ of masses $m_1, m_2, \ldots, m_n$, respectively. Assume that the rigid body is rotating about the axis with angular speed $\dot{\theta}$.

The diagram shows a plane section of the body, perpendicular to the axis and containing the particle P_i. The axis of rotation passes through the point O of this section and P_i is at a distance r_i from O.

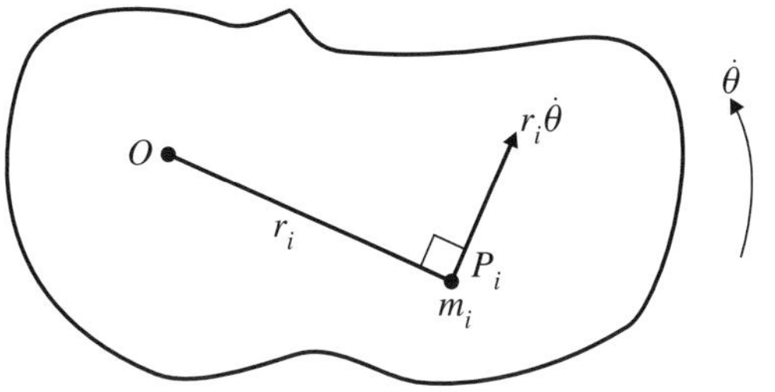

The particle P_i is moving in a circle of radius r_i and so has linear speed $r_i\dot{\theta}$ as shown. The linear momentum of P_i is therefore $m_i r_i \dot{\theta}$. As momentum is a vector, it has a moment about the axis through O which is calculated in the same manner as the moment of a force. So for the particle P_i, the moment of the momentum about the axis through O is given by

$$\text{moment of momentum} = (m_i r_i \dot{\theta}) r_i$$
$$= m_i r_i^2 \theta$$

Hence for the whole body,

$$\text{moment of momentum about axis through } O = \sum_i m_i r_i^2 \dot{\theta}$$
$$= \dot{\theta} \sum_i m_i r_i^2 = I\dot{\theta}$$

The moment of momentum is often called the **angular momentum** of the body.

When the mass is measured in kilograms, lengths in metres and angular speed in rad s^{-1}, the angular momentum is measured in $\text{kg m}^2\,\text{s}^{-1}$.

Example 6

A uniform rod AB of mass 0.5 kg and length 1.2 m has a particle of mass 1.5 kg attached to end B. The rod can rotate in a horizontal plane about a smooth vertical axis through A. Calculate the angular momentum of the rod and particle when the rod is rotating with angular speed $5\,\text{rad s}^{-1}$.

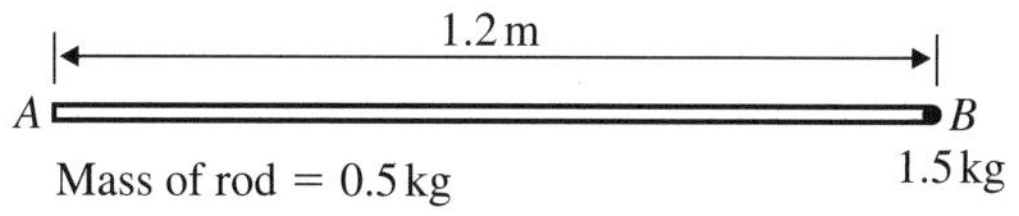

M.I. of rod about axis through mid-point perpendicular to rod $= \frac{1}{3} \times 0.5 \times 0.6^2\,\text{kg m}^2 = 0.06\,\text{kg m}^2$.

M.I. of rod about axis through A perpendicular to rod $= [0.06 + (0.5 \times 0.6^2)]\,\text{kg m}^2 = 0.24\,\text{kg m}^2$ (parallel axis theorem).

M.I. of particle about axis through A perpendicular to rod $= 1.5 \times 1.2^2\,\text{kg m}^2 = 2.16\,\text{kg m}^2$.

M.I. of rod and particle about axis through A perpendicular to rod $= (0.24 + 2.16)\,\text{kg}\,\text{m}^2 = 2.4\,\text{kg}\,\text{m}^2$.

$$\text{Angular speed} = 5\,\text{rad}\,\text{s}^{-1}$$

So
$$\begin{aligned}\text{angular momentum} &= I\dot{\theta}\\ &= 2.4 \times 5\,\text{kg}\,\text{m}^2\,\text{s}^{-1}\\ &= 12\,\text{kg}\,\text{m}^2\,\text{s}^{-1}\end{aligned}$$

4.5 Conservation of angular momentum

When any two bodies A and B collide, by Newton's third law they exert equal but opposite impulses on each other. If the motion of at least one of the bodies is rotational, then the moments of these impulses about the axis of rotation must also be equal but opposite.

Suppose A is rotating, and let the moment of the impulse on A from B be M. Then the moment of the impulse on B is $-M$. Provided there are no external forces which have a moment about the axis of rotation, the gain of angular momentum of A and B is zero. This is the **principle of conservation of angular momentum.**

- **Provided there are no external forces on a system with a moment about the axis of rotation, the total angular momentum of the system is constant.**

Example 7

A uniform rod AB of length 1.6 m and mass 3 kg is at rest on a smooth, horizontal table. AB is free to rotate on the table about a smooth, vertical axis through end A. A particle P of mass 0.6 kg is moving at $2\,\text{m}\,\text{s}^{-1}$ on the table at right angles to AB and strikes the rod at C where $AC = 0.9\,\text{m}$. P adheres to the rod. Calculate the angular speed with which the rod begins to rotate.

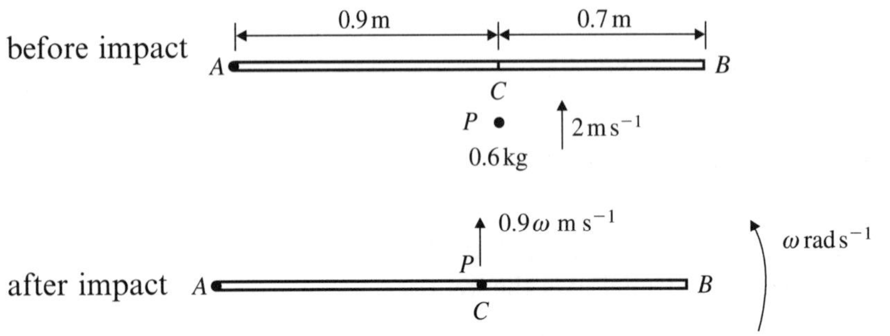

Let the rod begin to rotate with angular speed $\omega\,\text{rad}\,\text{s}^{-1}$.

M.I. of rod about axis through centre perpendicular to rod

$$= \tfrac{1}{3}ma^2 = \tfrac{1}{3} \times 3 \times \left(\frac{1.6}{2}\right)^2 \text{kg m}^2 = 0.64 \text{ kg m}^2$$

M.I. of rod about axis through A perpendicular to rod

$$= [0.64 + (3 \times 0.8^2)] \text{ kg m}^2 = 2.56 \text{ kg m}^2$$

Initial angular momentum of rod $= 0 \text{ kg m}^2 \text{ s}^{-1}$

Final angular momentum of rod about axis through A perpendicular to rod

$$= I\omega = 2.56\omega \text{ kg m}^2 \text{ s}^{-1}$$

Just before hitting the rod, P is moving at right angles to AB with a speed of 2 m s^{-1} at a distance 0.9 m from A.

So linear momentum of $P = 0.6 \times 2 \text{ N s} = 1.2 \text{ N s}$

Initial moment of momentum of P about vertical axis through A

$$= 1.2 \times 0.9 \text{ kg m}^2 \text{ s}^{-1}$$
$$= 1.08 \text{ kg m}^2 \text{ s}^{-1}$$

Just after the impact P is moving in a circle, radius 0.9 m, centre A, with a linear speed of $0.9\omega \text{ m s}^{-1}$.

So final moment of momentum of P about vertical axis through A

$$= mvr = 0.6 \times 0.9\omega \times 0.9 \text{ kg m}^2 \text{ s}^{-1}$$
$$= 0.486\omega \text{ kg m}^2 \text{ s}^{-1}$$

By the principle of conservation of angular momentum:

$$0 + 1.08 = 2.56\omega + 0.486\omega$$
$$\omega = \frac{1.08}{3.046} = 0.355$$

The rod begins to rotate with an angular speed of 0.355 rad s^{-1}.

4.6 The effect of an impulse on a rigid body which is free to rotate about a fixed axis

Consider a rigid body which is free to rotate about a fixed axis. The equation of rotational motion states:

$$L = I\ddot{\theta}$$

where $\ddot{\theta}$ is the angular acceleration of the body and L and I are the moment of the resultant force on the body and the moment of inertia of the body, respectively, about the axis of rotation.

Integrating with respect to time gives

$$\int_{t_1}^{t_2} L\,\mathrm{d}t = \int_{\omega_1}^{\omega_2} I\ddot{\theta}\,\mathrm{d}t$$

where ω_1 and ω_2 are the angular speeds at times t_1 and t_2 respectively.

$$\int_{t_1}^{t_2} L\,\mathrm{d}t = \left[I\dot{\theta}\right]_{\omega_1}^{\omega_2}$$

$$\int_{t_1}^{t_2} L\,\mathrm{d}t = I\omega_2 - I\omega_1 \qquad (1)$$

$\int_{t_1}^{t_2} L\,\mathrm{d}t$ is the **impulsive moment** of the resultant force about the axis and $(I\omega_2 - I\omega_1)$ is the gain of angular momentum. So equation (1) can be expressed in the form

- **impulsive moment of the force = gain of angular momentum**

Sometimes a rigid body is rotating under the action of a force of constant magnitude F acting at a constant distance r from the axis in a plane perpendicular to the axis. A cylinder rotating about its axis with a string wound around its circumference which is being pulled with a constant tension would be an example of this.

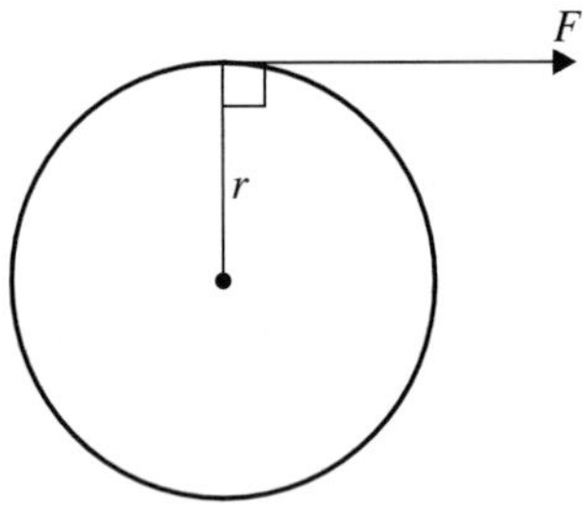

In this case, $L = F \times r$

So $$\int_{t_1}^{t_2} L\,\mathrm{d}t = \int_{t_1}^{t_2} F \times r\,\mathrm{d}t = Fr(t_2 - t_1) = F(t_2 - t_1)r$$

and hence:

impulsive moment of force = moment of force × time for which force acts

or impulsive moment = impulse × distance from axis

Impulsive moment is measured in N m s.

As impulsive moment equals change of angular momentum it follows that angular momentum also has units N m s. In other words, $\mathrm{kg\,m^2\,s^{-1}}$ and N m s are equivalent units.

Example 8

A uniform rod AB of mass 2 kg and length 0.9 m rests on a smooth, horizontal table and is free to rotate about a vertical axis through the end A. The rod receives a blow from a hammer at its free end in a direction perpendicular to the rod and begins to rotate at $10\,\mathrm{rad\,s^{-1}}$. Calculate the impulse of the blow.

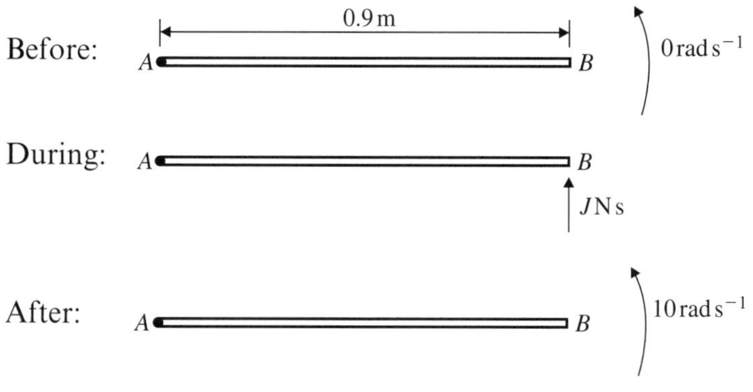

Let the impulse be J N s.

Moment of impulse about $A = 0.9J$ N m s

M.I. of rod about axis through centre perpendicular to rod

$$= \tfrac{1}{3}ma^2 = \tfrac{1}{3} \times 2 \times 0.45^2 \text{ kg m}^2 = 0.135 \text{ kg m}^2$$

M.I. of rod about axis through A perpendicular to rod

$$= [0.135 + (2 \times 0.45^2)] \text{ kg m}^2$$
$$= 0.54 \text{ kg m}^2$$

Initial angular momentum of rod $= 0$ N m s

Final angular momentum of rod $= I\omega$

$$= 0.54 \times 10 \text{ N m s}$$
$$= 5.4 \text{ N m s}$$

But moment of impulse = gain of angular momentum

so $$0.9J = 5.4$$

and $$J = \frac{5.4}{0.9} = 6$$

The impulse is 6 N s.

Example 9

A pulley of mass m and radius r is free to rotate in a vertical plane about a smooth, horizontal axis through its centre O. One end of a light inextensible string is attached to a point on the rim of the pulley. The string is wrapped several times around the pulley; a length of $5r$ remains free. A particle P of mass $2m$ is attached to the free end of the string. P is held close to the rim of the pulley, with OP horizontal, and is then released from rest.

(a) State a suitable model for the pulley.

(b) Find, in terms of r and g, the angular speed of the pulley immediately after the string becomes taut.

(a) The pulley can be modelled as a uniform disc.

(b)

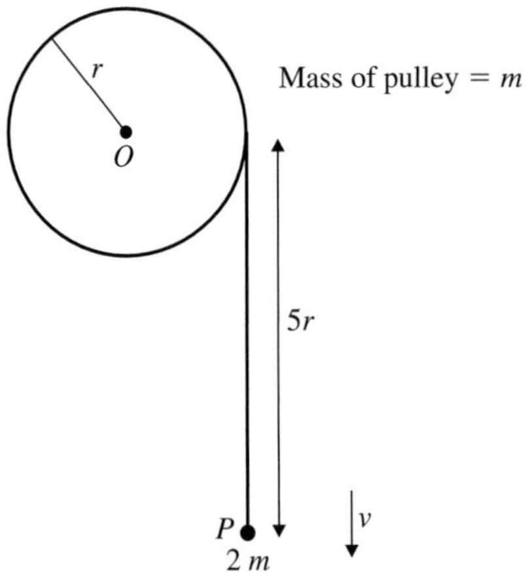

Let the speed of P immediately before the string becomes taut be v, and let the angular speed of the disc immediately after the string becomes taut be ω. P has then fallen a distance $5r$ from rest.

Using $v^2 = u^2 + 2as$ with $u = 0$, $a = g$, and $s = 5r$ gives

$$v^2 = 2 \times g \times 5r$$

$$v^2 = 10gr$$

$$v = \sqrt{(10gr)}$$

So the linear momentum of P just before the jerk as the string becomes taut is

$$2mv = 2m\sqrt{(10gr)}$$

Initial moment of momentum of $P = 2m\sqrt{(10gr)} \times r$

$$= 2mr\sqrt{(10gr)}$$

Final linear momentum of $P = 2mr\omega$

$\therefore$ final moment of momentum of $P = (2mr\omega)r = 2mr^2\omega$

M.I. of disc about axis through its centre $= \frac{1}{2}mr^2$

Initial angular momentum of disc $= 0$

Final angular momentum of disc $= I\omega = \frac{1}{2}mr^2\omega$

By the principle of conservation of angular momentum:

$$0 + 2mr\sqrt{(10gr)} = \tfrac{1}{2}mr^2\omega + 2mr^2\omega$$

$$2\sqrt{(10gr)} = \tfrac{5}{2}r\omega$$

$$\omega = \tfrac{4}{5}\sqrt{\left(\frac{10g}{r}\right)}$$

The angular speed of the pulley is $\frac{4}{5}\sqrt{\left(\frac{10g}{r}\right)}$.

Exercise 4C

1 A uniform disc of mass 1.5 kg and radius 2 m is rotating at a constant angular speed of $5\,\text{rad}\,\text{s}^{-1}$ about a perpendicular axis through its centre. Calculate the angular momentum of the disc.

2 A uniform square lamina of mass 2 kg and side 0.5 m is free to rotate about an axis which coincides with one of its sides. Calculate the loss of angular momentum when the angular speed of the lamina is reduced from $4\,\text{rad}\,\text{s}^{-1}$ to $1\,\text{rad}\,\text{s}^{-1}$.

3 A uniform rod AB of length $4a$ and mass m is free to rotate in a vertical plane about a smooth, fixed, horizontal axis through the point C of the rod where $AC = a$. The rod is released from rest with AB horizontal. When the rod first becomes vertical, end B strikes a stationary particle of mass m which adheres to the rod. Find, in terms of g and a, the angular speed of the rod after the impact and calculate the angle between the rod and the downward vertical when the rod comes to instantaneous rest.

4 A rectangular sign is hanging outside a shop. The sign has mass 2 kg and measures 1 m by 2 m. It is free to swing about a smooth, fixed, horizontal axis which coincides with a long side of the rectangle. The sign is initially hanging at rest when it receives an impulse at its centre of mass in a direction perpendicular to its plane. The sign first comes to rest in a horizontal position. By assuming that the sign can be modelled as a uniform rectangular lamina, calculate the initial angular speed of the sign and the magnitude of the impulse.

5 A uniform rod AB of length $2a$ and mass m is free to rotate in a vertical plane about a fixed, smooth, horizontal axis through A. The rod is initially at rest with B vertically below A. The end B receives a horizontal impulse of magnitude J in a direction perpendicular to the axis of rotation. Find, in terms of m, g and a, the least value of J if the rod is to rotate in complete circles. Given that the magnitude of J is one half of this value, find the angle the rod turns through before first coming to instantaneous rest.

6 A uniform rod AB of mass 0.2 kg and length 0.8 m is free to rotate in a vertical plane about a fixed, smooth, horizontal axis through its centre. It is initially at rest in a horizontal position. A particle of mass 0.15 kg falls vertically onto the rod, and sticks to the rod at the point C where $AC = 0.1$ m. The rod subsequently makes complete revolutions. Calculate the minimum height above the rod from which the particle must have fallen.

7 A uniform solid cylindrical drum of mass 1.5 kg and radius 0.5 m is free to rotate about a fixed, smooth, horizontal axis which coincides with the axis of the cylinder. The axis is at a height of 2 m above a horizontal table, and a light string AB of length 4 m has one end attached to the highest point of the cylinder. A block of mass 0.3 kg is attached to end B of the string and rests on the table, as shown in the diagram. The drum begins to rotate at a constant angular speed of 4 rad s^{-1} in a clockwise direction. Calculate the angular speed of the drum immediately after the block is jerked into motion.

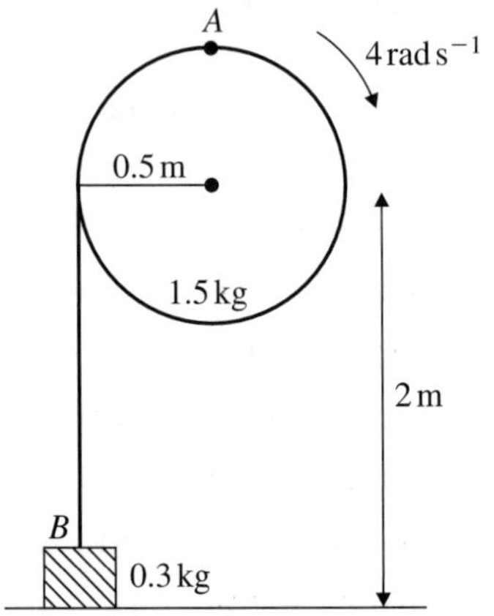

8 A uniform rod AB of mass m and length $6a$ is free to rotate about a fixed, smooth, vertical axis through the point C of AB, where $AC = 2a$. AB is at rest on a smooth, horizontal table when it is struck by a particle of mass $2m$ moving on the table in a direction perpendicular to the rod with speed u. The particle adheres to the rod at D, where $DB = x$. Given that the speed of the particle is halved by the impact, find x.

9 A uniform rod AB of mass m and length $2a$ is free to rotate about a fixed, smooth, vertical axis through A. AB is at rest on a smooth, horizontal table. A particle P of mass $6m$ moving in a direction perpendicular to the rod with speed u hits the rod at C where $AC = \dfrac{3a}{4}$. P is brought to rest by the impact. Find the angular speed of the rod after the impact.

10 A uniform square lamina of mass m and side l is free to rotate about a fixed, smooth, horizontal axis which coincides with a side of the lamina. The lamina is hanging in equilibrium when it is hit at its centre of mass by a particle of mass $3m$ moving with speed v in a direction perpendicular to the plane of the lamina. The particle adheres to the lamina. Find, in terms of v and l, the angular speed of the lamina immediately after the impact. Hence show that the lamina will perform complete revolutions if

$$v^2 > \tfrac{104}{27}lg$$

11 Four rods, each of mass $3m$ and length $2l$, are joined together to form a square framework $ABCD$. The framework is free to rotate in its own plane about a smooth, fixed, horizontal axis through A perpendicular to the plane of the framework. The framework is hanging in equilibrium with C vertically below A when a horizontal impulse in the plane of the framework is applied at the mid-point of AB. The framework first comes to rest when AC makes an angle $\frac{\pi}{4}$ with the upward vertical. Find the magnitude of the impulse.
If instead the impulse were applied at the mid-point of BC, determine whether the framework would perform complete revolutions.

12 A uniform square lamina of side $2l$ and mass m is rotating with angular speed ω in a horizontal plane about a smooth fixed vertical axis through the centre of the lamina. A particle of mass $3m$ is held at a height l above the lamina. The particle is released and adheres to the lamina at a corner. Find the new angular speed of the lamina.

13 A uniform rod AB of mass m and length $2a$ is free to rotate in a vertical plane about a fixed, smooth horizontal axis through point C of the rod, where $AC = \frac{1}{2}a$. The rod is hanging in equilibrium with B below A when it receives a horizontal impulse at B. The impulse is just sufficient to cause the rod to perform complete revolutions. Find, in terms of m, a and g, the magnitude of the impulse.

14 A light inextensible string has one end attached to the rim of a pulley of mass $3m$ and radius r and is wound several times around the pulley. A pan of mass m is attached to the other end of the string and hangs freely below the pulley. The system is held at rest. A particle of mass $5m$ is dropped from rest from a height $4r$ above the pan. The pulley system is released from rest at the moment the particle hits the pan. The particle adheres to the pan. Assuming that the pulley can be modelled as a uniform circular disc and the pan as a particle, find an expression for the angular speed of the pulley immediately after the impact.

15 A pulley in the form of a uniform disc of mass $4m$ has centre O and radius a. It is free to rotate in a vertical plane about a fixed smooth horizontal axis through O. A light inextensible string has one end attached to a point on the rim of the pulley and is wrapped several times round the rim. The length of the string not wrapped round the rim is $8a$ and has a particle P, of mass m, attached at its free end. Initially the disc is at rest and P is held close to the rim of the disc and level with O. The particle P is now released. Determine the angular speed of the disc immediately after the string becomes taut.

4.7 The simple pendulum

A **simple pendulum** consists of a particle, P, sometimes called a bob, attached to one end of a light string, the other end of which is fixed. The pendulum swings through small angles on either side of the vertical.

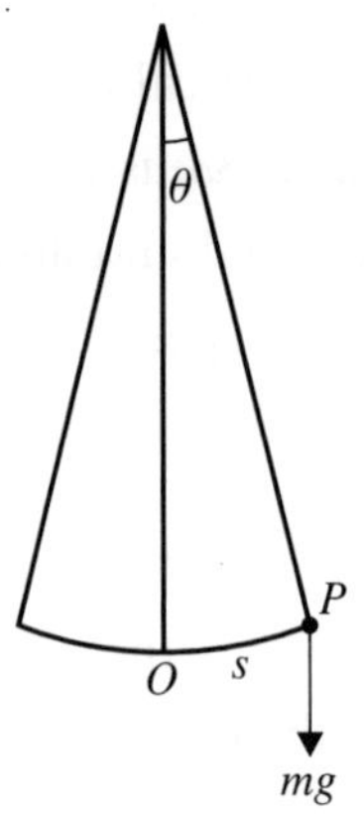

Let the length of the string be l.

Consider the situation when the particle is at a distance s from the centre O of the oscillation, where s is measured along the arc that the particle describes.

Then $$s = l\theta$$

Differentiating with respect to time gives

$$\dot{s} = l\dot{\theta}$$

and $$\ddot{s} = l\ddot{\theta}$$

The component of the force acting on the particle in the direction of the tangent to the arc described is $mg\sin\theta$ *towards* the centre of oscillation, O. So by Newton's second law:

$$F = ma$$

$$mg\sin\theta = -m\ddot{s}$$

or $$mg\sin\theta = -ml\ddot{\theta}$$

But for small angles: $$\sin\theta \simeq \theta$$

(This approximation can be obtained from the series expansion for $\sin\theta$ (see Book P6). Alternatively use your calculator to convince yourself of the validity of this statement.)

And so $$mg\theta = -ml\ddot{\theta}$$

$$\ddot{\theta} = -\frac{g}{l}\theta$$

Comparing this equation with the standard simple harmonic motion (S.H.M.) equation $\ddot{x} = -\omega^2 x$ shows that the pendulum is moving with S.H.M. of period $2\pi\sqrt{\left(\frac{l}{g}\right)}$.

The calculation above was made using several assumptions. The bob was assumed to be a particle, the string was assumed to be light and inelastic, the point of support was assumed to be fixed and the amplitude of the oscillation was small so that $\sin\theta \simeq \theta$. A simple pendulum model is valid for a body swinging on the end of a string or rod, the other end of which is fixed, only if these assumptions are reasonable. In particular:

(i) The mass of the body must be large compared with the weight of the string or rod, otherwise the string cannot be assumed light.
(ii) The size of the body must be small compared with the length of the string, or the body cannot be assumed to be a particle.
(iii) The amplitude of the oscillation must be small. The formula for the period has been found to be very accurate for amplitudes up to 0.26^c (approximately 15°) but thereafter the accuracy diminishes.

Provided these assumptions are valid for a particular pendulum arrangement, the simple pendulum model is valid. The period of oscillation, T, is then independent of the mass of the bob and is given by

■ $$\boldsymbol{T = 2\pi\sqrt{\left(\frac{l}{g}\right)}}$$

where l is the length of the string or rod.

Example 10

A simple pendulum is found to have a period of 2 seconds. Calculate the length of the string.

$$\text{Period} = 2\pi\sqrt{\left(\frac{l}{g}\right)}$$

so

$$2 = 2\pi\sqrt{\left(\frac{l}{9.8}\right)}$$

$$\sqrt{\left(\frac{l}{9.8}\right)} = \frac{2}{2\pi} = \frac{1}{\pi}$$

Squaring both sides:

$$\frac{l}{9.8} = \frac{1}{\pi^2}$$

$$l = \frac{9.8}{\pi^2} = 0.9929$$

The length of the pendulum is 0.993 m.

4.8 The compound pendulum

If a body is making small oscillations about a fixed axis but the body does not satisfy the criteria to be modelled as a simple pendulum, then a refined model is required.

Consider a rigid body which can rotate about a smooth, fixed, horizontal axis and is initially hanging in equilibrium. If it is slightly displaced, it will swing like a pendulum. Such a swinging body is called a **compound pendulum**. The diagram shows the vertical plane through the centre of mass, G, of the body. The axis passes through point O of this plane.

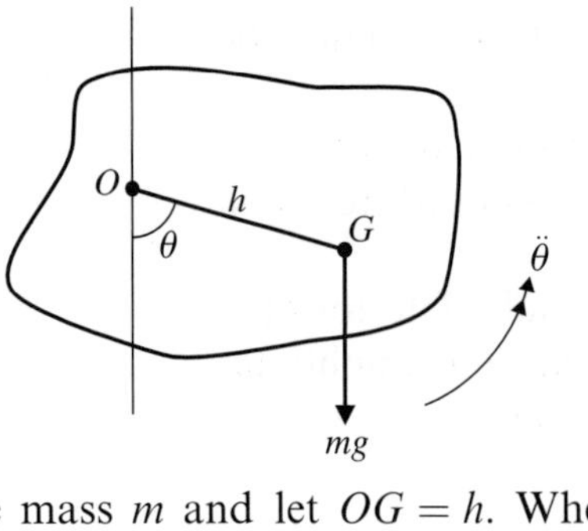

Let the body have mass m and let $OG = h$. When OG makes an angle θ with the downward vertical, the angular acceleration of the body is $\ddot{\theta}$, as shown. Let the moment of inertia of the body about the given axis be I.

The moment of the weight of the body about the axis is $mgh\sin\theta$ ↻. As this is the only force with a non-zero moment about the axis, the equation of rotational motion gives

$$\text{moment of resultant force} = I\ddot{\theta} \circlearrowleft$$

so

$$mgh \sin\theta = -I\ddot{\theta} \circlearrowright$$

For small oscillations, as with the simple pendulum, $\sin\theta \approx \theta$ and so

$$\ddot{\theta} = \frac{-mgh\theta}{I}$$

Comparing this with the standard S.H.M. equation $\ddot{x} = -\omega^2 x$ shows that we have, approximately, an oscillation of period $2\pi\sqrt{\left(\frac{I}{mgh}\right)}$.

For a rigid body of mass m with radius of gyration k about a parallel axis through its centre of mass, G, the moment of inertia about a horizontal axis through G is mk^2. Hence, by the parallel axis rule, the moment of inertia I about the horizontal axis through O is given by

$$I = mk^2 + mh^2 = m(k^2 + h^2)$$

The period of small oscillations about the axis through O for this body is now given by

$$T = 2\pi\sqrt{\left(\frac{I}{mgh}\right)}$$

$$= 2\pi\sqrt{\left[\frac{m(k^2 + h^2)}{mgh}\right]}$$

■ **So** $$\boldsymbol{T = 2\pi\sqrt{\left[\frac{k^2 + h^2}{gh}\right]}}$$

where h is the distance of the centre of mass, G, from the axis and k is the radius of gyration of the body about a parallel axis through G.

Example 11

A uniform rod AB of length $2a$ can swing in a vertical plane about a smooth, horizontal axis through point C of the rod where $AC = \frac{a}{2}$. Calculate the period of small oscillations of the rod about its equilibrium position.

Let the rod have mass m.

M.I. of rod about axis through centre, perpendicular to rod

$$= \tfrac{1}{3}ma^2$$

Radius of gyration, k, about same axis is given by

$$k^2 = \tfrac{1}{3}a^2$$

$$\text{Period of small oscillations} = 2\pi\sqrt{\left[\frac{k^2+h^2}{gh}\right]}$$

$$= 2\pi\sqrt{\left[\frac{\frac{1}{3}a^2 + \frac{a^2}{4}}{g\frac{a}{2}}\right]}$$

$$= 2\pi\sqrt{\left[\frac{\frac{7a^2}{12}}{\frac{ga}{2}}\right]}$$

$$= 2\pi\sqrt{\left[\frac{7a}{6g}\right]}$$

Example 12

A thin uniform rod AB, of mass m and length $2a$, is free to rotate in a vertical plane about a fixed, smooth, horizontal axis through A. A uniform circular disc, of mass $3m$ and radius $\frac{a}{3}$, is clamped to the rod with its centre O on the rod so that it lies in the plane of rotation. If $OA = x$ show that the moment of inertia of the system about the axis is

$$\frac{3m}{2}(a^2 + 2x^2)$$

The system is initially hanging in equilibrium with B vertically below A. It is then slightly displaced and oscillates freely under gravity. Find the period of small oscillations and show that this is least when

$$x = \frac{a}{6}(\sqrt{22} - 2)$$

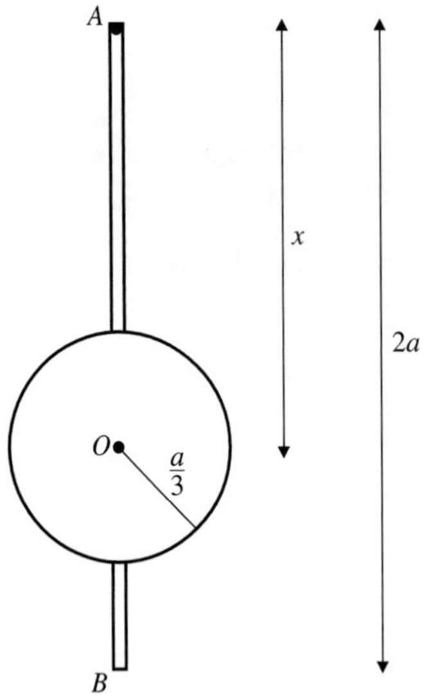

M.I. of rod about horizontal axis through mid-point

$$= \tfrac{1}{3}ma^2$$

M.I. of rod about horizontal axis through A

$$= \tfrac{1}{3}ma^2 + ma^2 = \tfrac{4}{3}ma^2$$

M.I. of disc about perpendicular axis through O

$$= \tfrac{1}{2} \times 3m \times \left(\frac{a}{3}\right)^2 = \frac{ma^2}{6}$$

M.I. of disc about parallel axis through A

$$= \frac{ma^2}{6} + 3mx^2$$

So M.I. of system about the given axis through A

$$= \frac{4ma^2}{3} + \frac{ma^2}{6} + 3mx^2$$

$$= \frac{9ma^2}{6} + 3mx^2$$

$$= \frac{3m}{2}(a^2 + 2x^2)$$

As the moment of inertia of the system about the given axis is now known, use the formula

$$T = 2\pi\sqrt{\left(\frac{I}{Mgh}\right)}$$

to find the period of small oscillations.

h is the distance of the centre of mass from the axis. So you must find the position of the centre of mass.

	Rod	Disc	Composite body
Mass	m	$3m$	$4m$
Distance of centre of mass from A	a	x	h

Taking moments about the axis of rotation gives

$$ma + 3mx = 4mh$$

so

$$h = \frac{a + 3x}{4}$$

Hence:

$$T = 2\pi\sqrt{\left(\frac{I}{Mgh}\right)}$$

where $I = \dfrac{3m}{2}(a^2 + 2x^2)$, $M = 4m$ and $h = \dfrac{a + 3x}{4}$.

So:
$$T = 2\pi\sqrt{\left[\frac{3m(a^2+2x^2)\times 4}{2\times 4mg(a+3x)}\right]}$$
$$= 2\pi\sqrt{\left[\frac{3(a^2+2x^2)}{2(a+3x)g}\right]}$$

The period will be least when $\dfrac{(a^2+2x^2)}{(a+3x)}$ is least.

Let $f(x) = \dfrac{(a^2+2x^2)}{(a+3x)}$.

Then $f(x)$ is least when $\dfrac{df(x)}{dx} = 0$.

$$\frac{df(x)}{dx} = \frac{4x(a+3x)-(a^2+2x^2)\times 3}{(a+3x)^2}$$

$\dfrac{df(x)}{dx} = 0$ when

$$4x(a+3x)-(a^2+2x^2)\times 3 = 0$$
$$4ax+12x^2-3a^2-6x^2 = 0$$
$$6x^2+4ax-3a^2 = 0$$

So:
$$x = \frac{-4a\pm\sqrt{(16a^2+72a^2)}}{12}$$
$$= \frac{-4a\pm\sqrt{(88a^2)}}{12}$$
$$= \frac{-4a\pm 2a\sqrt{22}}{12}$$
$$= \frac{-2a\pm a\sqrt{22}}{6}$$

As x must be positive,

$$x = \frac{-2a+a\sqrt{22}}{6}$$
$$= \frac{a}{6}(\sqrt{22}-2)$$

The period of small oscillations is least when $x = \frac{a}{6}(\sqrt{22}-2)$.

The equivalent simple pendulum

The simple pendulum which has the same period as a given compound pendulum is called the **equivalent simple pendulum**.

The length of the equivalent simple pendulum can be found by recalling that the period of a simple pendulum of length l is $2\pi\sqrt{\left(\frac{l}{g}\right)}$. From above, the period of the compound pendulum is $2\pi\sqrt{\left(\frac{I}{mgh}\right)}$. Thus the equivalent simple pendulum has length $\frac{I}{mh}$.

Alternatively, for a rigid body with radius of gyration k about a parallel axis through its centre of mass G,

$$T = 2\pi\sqrt{\left[\frac{k^2 + h^2}{gh}\right]}$$

and the equivalent simple pendulum has length

$$l = \frac{k^2 + h^2}{h}$$

Example 13

A uniform disc of radius a is performing small oscillations in a vertical plane about a smooth, horizontal axis which is perpendicular to the disc and passes through a point at a distance $\frac{a}{2}$ from its centre. Find the length of the equivalent simple pendulum.

M.I. of disc about axis through its centre $= \frac{1}{2}ma^2$

So M.I. of disc about axis $\frac{a}{2}$ from its centre

$$= \tfrac{1}{2}ma^2 + m\left(\frac{a}{2}\right)^2$$

$$= \frac{3ma^2}{4}$$

and length of equivalent simple pendulum $= \dfrac{I}{mh}$

$$= \frac{3ma^2}{4} \times \frac{1}{m \times \frac{a}{2}}$$

$$= \frac{3a}{2}$$

Exercise 4D

Whenever a numerical value of g is required, take $g = 9.8\,\text{m s}^{-2}$.

1 A simple pendulum is performing small oscillations. Calculate the period of the pendulum given that the length is (a) 0.5 m, (b) 1.3 m, (c) 70 cm.

2 A simple pendulum is performing small oscillations. Calculate the length of the pendulum, given that the period is

(a) $\frac{3\pi}{8}$ s (b) $\frac{4\pi}{9}$ s (c) 1 s

3 A simple pendulum takes one second to perform half a complete oscillation. Calculate the length of string required for the pendulum. (A pendulum with this period is called a seconds pendulum.)

4 A simple pendulum has length L and period T. If the length is reduced to $\frac{1}{2}L$, calculate the new period in terms of T.

5 A simple pendulum has period T and length L. The period is to be increased to $3T$. Calculate the new length in terms of L.

6 A plumb line consists of a light cord of length 2 m with a metal bob attached to one end. The other end of the cord is fixed and the bob is hanging freely. The bob is slightly displaced and the plumb line is seen to oscillate. State the assumptions you would be making about the cord, the metal bob and the oscillations if you were to model the plumb line as a simple pendulum. Assuming the simple pendulum model to be a valid one, calculate the period of the oscillations.

7 A boy is swinging on the end of a rope. The other end of the rope is tied to a branch of a tree. What assumptions must be made if his motion is to be considered to be that of a simple pendulum? Given that this model is valid and the period of oscillation is 3.5 s, calculate the length of the rope.

Calculate
(i) the period of small oscillations about their equilibrium positions,
(ii) the length of the simple equivalent pendulum,
for each of the bodies in questions **8–13** with rotation about the given axes.

8 A uniform rod of mass m and length $2a$, axis through one end, perpendicular to the rod.

9 A uniform rod of mass m and length $2a$ with a particle of mass $3m$ attached at one end, axis through the other end, perpendicular to the rod.

10 A uniform circular hoop of mass $2m$ and diameter d, axis through a point on the hoop, perpendicular to the plane of the hoop.

11 A triangular framework formed by joining three uniform rods each of length $2a$ and mass m at their ends, axis through a vertex of the triangle perpendicular to the plane of the triangle.

12 A uniform rectangular lamina of mass 3 kg measuring 2 m by 4 m, axis coinciding with a short side of the lamina.

13 A uniform circular disc, centre O, mass m and radius r, with a particle mass m attached at A where $OA = \frac{1}{2}r$, axis through O perpendicular to the disc.

14 A uniform rod AB of length $2a$ and mass m is free to rotate about a fixed, smooth, horizontal axis through A. Find the period of small oscillations of the rod. A particle is now attached to end B of the rod and as a result the period of small oscillations is increased by 20%. Find the mass of the particle.

15 A pendulum consists of a uniform rod, of mass M and length $2a$, pivoted at its mid-point O and a small nut of mass m which can be screwed to any point of the rod.
(a) Find the period T of a small oscillation in a vertical plane when the nut is a distance x from O.
(b) Show that if $3m < M$ then T decreases as x increases.
(c) When $3m > M$ find the value of x which makes T a minimum.

16 A thin uniform rod AB, of mass m and length $6a$, can turn freely in a vertical plane about a smooth, horizontal axis through end A. A uniform circular disc of mass $12m$ and radius a is clamped to the rod so that its plane coincides with the plane in which the rod can turn.
(a) Show that for different positions of the disc the minimum length of the equivalent simple pendulum is $2a$.
(b) Find the maximum length of the equivalent simple pendulum.

17

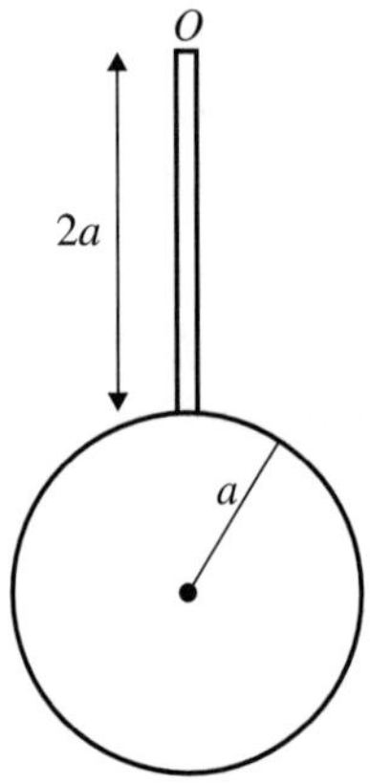

The figure shows a compound pendulum which consists of a uniform solid sphere of mass m and radius a attached to a uniform rod of mass $\frac{1}{4}m$ and length $2a$. They are attached so that the centre of the sphere lies on the rod produced and so is a distance $3a$ from the pivot O. Find the moment of inertia of the system about a horizontal axis through O and hence the length of the equivalent simple pendulum.

18

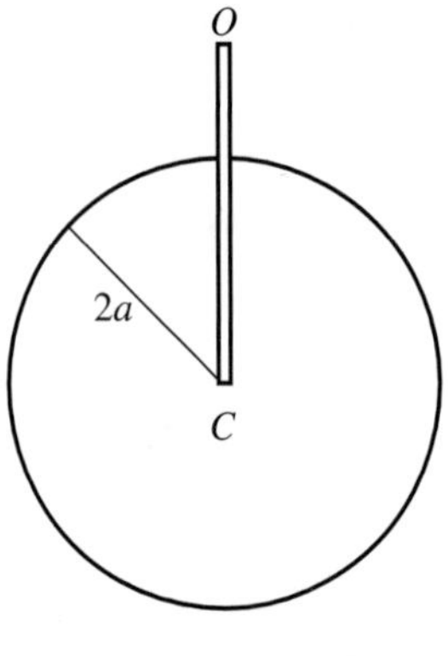

The figure shows a compound pendulum consisting of a thin uniform rod OC, of length $3a$ and mass M, rigidly attached at C to the centre of a uniform disc of radius $2a$ and mass M. The rod OC is in the same vertical plane as the disc. The pendulum is free to rotate in this vertical plane about a fixed smooth horizontal axis through O perpendicular to the plane of the disc.

(a) Show that the moment of inertia of the pendulum about the axis through O is $14Ma^2$.

Given that the pendulum performs small oscillations about its position of stable equilibrium,

(b) find, in terms of g and a, the period of oscillation.

19 An ear-ring of mass m is formed by cutting out a circle of radius a from a thin uniform circular disc of metal of radius $3a$, as shown in the diagram. The centre A of the larger disc, the centre B of the smaller circle and the point X on the circumference of both circles are collinear. The ear-ring is free to rotate about a smooth horizontal axis through X perpendicular to the plane of the ear-ring. Show that the period of small oscillations about the position of stable equilibrium is $4\pi\sqrt{\frac{15a}{13g}}$.

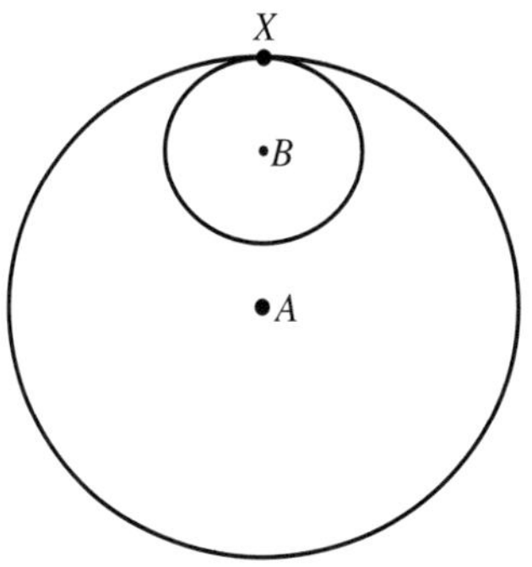

20 A uniform rod AB of length $4l$ and mass $6m$ has a uniform solid sphere attached to each end. The centres of both spheres lie on the same line as the rod. Each sphere has radius $\frac{l}{2}$ and mass m. The rod is free to rotate about a fixed smooth horizontal axis through point C of the rod, where $AC = l$. The axis is perpendicular to the rod. Find the period of small oscillations of the system about its position of stable equilibrium.

21

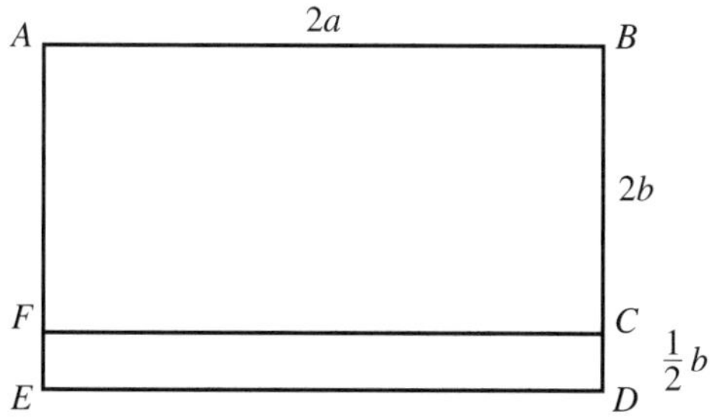

The diagram shows a rectangular sign outside a shop. The sign is composed of two portions, both of which are rectangular. $ABCF$ has mass m, length $2a$ and breadth $2b$. $FCDE$ has mass $2m$, length $2a$ and breadth $\frac{1}{2}b$. The sign is free to rotate about a horizontal axis along AB. The wind causes it to make small oscillations about its position of stable equilibrium. Find an expression for the period of these oscillations.

SUMMARY OF KEY POINTS

1 The kinetic energy of a rigid body which is rotating with angular speed $\dot{\theta}$ is given by

$$\text{K.E.} = \tfrac{1}{2}I\dot{\theta}^2$$

where I is the moment of inertia of the body about the axis of rotation.

2 When a rigid body is rotating, the gain in potential energy of the body is given by

$$\text{Gain in P.E.} = mgh$$

where m is the mass of the body and h is the vertical height gained by its centre of mass.

3 When a rigid body is rotating about a smooth axis with no external forces acting on the body, the sum of the kinetic and potential energies of the body will remain constant.

4 The equation of rotational motion states:

$$L = I\ddot{\theta}$$

where L is the moment about the axis of rotation of the resultant force on the rotating body, I is the moment of inertia of the body about the axis of rotation and $\ddot{\theta}$ is the angular acceleration of the body.

5 For a rigid body which is rotating about a fixed, smooth axis, the force exerted by the axis on the body can be calculated by considering the motion of a particle of the same mass as the body placed at the centre of mass of the body under the action of the same forces as those acting on the body.

6 The angular momentum of a rigid body rotating with angular speed $\dot{\theta}$ is given by

$$\text{angular momentum} = I\dot{\theta}$$

7 When there are no external forces with moments about the axis of rotation acting on a system, the total angular momentum of the system remains constant.

8 The period T of a simple pendulum is given by

$$T = 2\pi\sqrt{\left(\frac{l}{g}\right)}$$

where l is the length of the pendulum.

9 For a rigid body performing small oscillations about a fixed, horizontal axis (a compound pendulum) the period of oscillations is

$$2\pi\sqrt{\left(\frac{I}{mgh}\right)}$$

where I is the moment of inertia of the body about the axis, m is the mass of the body and h is the distance of the centre of mass of the body from the axis.

10 For a rigid body performing small oscillations about a fixed horizontal axis, the period of oscillations is

$$2\pi\sqrt{\left[\frac{k^2+h^2}{gh}\right]}$$

where h is the distance of the centre of mass from the axis and k is the radius of gyration of the body about a parallel axis through the centre of mass.

11 The simple pendulum which has the same period as a given compound pendulum is called the equivalent simple pendulum. It has length

$$\frac{I}{mh} \quad \text{or} \quad \frac{k^2+h^2}{h}$$

Review exercise 2

1 Given that the moment of inertia of a uniform disc, of radius a and mass m, about an axis through its centre perpendicular to its plane is $\frac{1}{2}ma^2$, obtain the moment of inertia about a diameter of the disc.
Hence obtain the moment of inertia about a tangent to the disc in the plane of the disc. [E]

2 Show, by integration, that the moment of inertia of a uniform rod, of length $2l$ and mass m, about an axis through its centre and perpendicular to the rod is $\frac{1}{3}ml^2$.
A uniform square plate, of mass M, has edges of length $2a$. Find the moment of inertia of this plate about an axis through its centre perpendicular to the plane of the plate. [E]

3 A uniform circular disc of radius $2a$ has a concentric hole of radius a. The mass of the annulus is M. Find the moment of inertia of the annulus about an axis through a point on the circumference of the circle of radius a

(a) if the axis is perpendicular to the plane of the annulus,

(b) if the axis lies in the plane of the annulus and touches the circle. [E]

4 A uniform circular plate of diameter 5 m has a hole of diameter 1 m punched in it. The centre of the hole is 1.5 m from the centre of the plate. Calculate the radius of gyration of the plate

(a) about the diameter through the centre of the hole,

(b) about the diameter which is perpendicular to this.

5 A uniform square lamina $ABCD$ of mass m and side $2a$ is free to rotate in a vertical plane about an axis through its centre O perpendicular to the plane of the lamina. Particles each of mass m are attached to points A and B of the lamina. The system is released from rest with AB vertical. Show that the angular speed of the system when AB is horizontal is $\sqrt{\left(\frac{6g}{7a}\right)}$.

6 A uniform disc, of mass $2m$ and radius r, is free to rotate about a smooth horizontal axis passing through the centre of the disc and perpendicular to the plane of the disc. A light inextensible string, attached to the rim of the disc, passes round part of the rim and then vertically downwards to a hanging particle P of mass $3m$, to which the other end of the string is attached. The system starts from rest. Find the distance that P has fallen when it has acquired a speed V, assuming that some of the string is still wound round the rim. Find also the acceleration of P and the tension in that part of the string which is hanging vertically. [E]

7 A uniform circular disc of mass 12 kg and radius 10 cm is free to rotate about a horizontal axis through its centre perpendicular to its plane. A particle of mass 3 kg is attached to the highest point of the rim of the disc and the equilibrium of the system is slightly disturbed. Find the angular speed of the particle, in revolutions per minute, when the particle is passing through its lowest point. [E]

8 A wheel has a cord of length 10 m coiled round its axle; the cord is pulled with a constant force of 100 N and when the cord leaves the axle the wheel is rotating 5 times a second. Calculate the moment of inertia of the wheel and axle. [E]

9 Three equal uniform rods, each of length l and mass m, form the sides of an equilateral triangle ABC. The triangular frame is attached to a smooth hinge at A, about which it can rotate in a vertical plane. The frame is held, with AB horizontal and C below AB, and then let go from rest. Find the maximum angular speed of the triangle in the subsequent motion. [E]

10 A bucket of mass m hangs at the end of a light rope which is coiled round a wheel of mass M. If the wheel can rotate freely about its axis, which is horizontal, and if its entire mass is supposed concentrated in its rim, find the speed of the bucket when it has fallen a distance x from rest. [E]

11 A uniform wire of mass M and length $4a$ is bent in the form of a square. It is hung by one corner over a smooth horizontal nail and released from rest from the position in which the sides are horizontal and vertical and the centre of mass is at a lower level than the nail. Find the angular speed with which it passes through the position in which the diagonals are horizontal and vertical, and the resultant reaction on the nail as it passes through this position. [E]

12 Find the moment of inertia of a uniform square lamina $ABCD$, of side $2a$ and mass m, about an axis through A perpendicular to the plane of the lamina.
The lamina is free to rotate about a fixed smooth horizontal axis through A perpendicular to the plane of the lamina. Show that its period of small oscillations about the stable equilibrium position is

$$2\pi\left(\frac{8a}{3g\sqrt{2}}\right)^{\frac{1}{2}}$$

The lamina is rotating with angular speed ω when C is vertically below A. Determine the components, along and perpendicular to AC, of the reaction of the lamina on the axis when AC makes an angle θ with the downward vertical through A. [E]

13 A thin uniform rod, of length $2a$ and mass M, attached to a smooth fixed hinge at one end O, is allowed to fall from a horizontal position. Show that in the subsequent motion

$$2a\left(\frac{d\theta}{dt}\right)^2 = 3g\sin\theta$$

where θ is the angle made by the rod with the horizontal. Find, in terms of M, g and θ, the resolved parts, along the rod and perpendicular to the rod, of the force exerted by the rod on the axis of rotation.

Show that the resolved part of this force in a horizontal direction is greatest when $\theta = \frac{\pi}{4}$ and that the resolved vertical part is then $\dfrac{11Mg}{8}$. [E]

14 A uniform rod, of length $4a$ and mass m, is held at the edge of a horizontal table, with a length a resting on the table at right angles to the edge, and the remainder projecting beyond the table. The rod is released. Assuming that the edge of the table is rough enough to prevent slipping during rotation of the rod through an angle θ, find, in terms of m, g and θ, the reactions on the rod at the point of contact along and perpendicular to the rod.
Show that, if μ is the coefficient of friction, the rod will begin to slip when it has turned through an angle α given by

$$\tan \alpha = \frac{4\mu}{13}$$ [E]

15 Four uniform rods, each of mass m and length $2l$, are joined rigidly together to form a square frame $ABCD$ of side $2l$. The frame is placed with all four sides at rest on a smooth horizontal table. An inextensible string has one end attached at the corner A. A particle of mass $4m$ is tied to the other end of the string. The particle, initially at A, is projected with speed u in the direction DA. Given that the speed of the particle immediately after the string becomes taut is V, show that the initial angular speed of the square frame about an axis through its centre of gravity perpendicular to the plane of the frame is ω where $\omega = \dfrac{2V - u}{l}$. Show that $V = \dfrac{7u}{11}$, and that immediately after the string becomes taut the kinetic energy of the particle and frame is $\dfrac{14mu^2}{11}$. [E]

16 A uniform rigid rod AB, of mass M and length $2a$, is falling freely without rotation under gravity, with AB horizontal. Suddenly the end A is fixed when the speed of the rod is v. Find the angular speed with which the rod begins to rotate.

[E]

17 Find, by integration, the moment of inertia of a uniform circular disc, of mass M and radius a, about an axis which passes through the centre and is perpendicular to the plane of the disc. Hence show that the moment of inertia of such a disc about a tangent line in its plane is $\frac{5Ma^2}{4}$.

The disc rotates about a fixed smooth horizontal axis which is the tangent to the disc at the point A. The centre O of the disc describes a vertical circle with centre A in a plane perpendicular to this tangent. The disc is released from rest when its plane is horizontal. Find the angular speed of the disc when it is first vertical.

At the instant when the disc is in this vertical position, it is hit at its centre by a particle of mass M, travelling with speed u in the direction of motion of the centre of the disc. Given that the particle adheres to the disc, find, in terms of u, a and g, the angular speed of the system immediately after impact. [E]

18 A rod AB, of length $2a$ and mass $2m$, lies at rest on a smooth horizontal table, and is pivoted about a smooth vertical axis through A. A small body of mass m, moving on the table with speed V at right-angles to the rod, strikes the rod at a distance d from A. Given that the body sticks to the rod after impact, find the angular speed with which the rod starts to move. [E]

19 To the end B of a thin uniform rod AB, of length $3a$ and mass m, is attached a thin uniform circular disc, of radius a and mass m, so that the rod and the diameter BC of the disc are in a straight line with $AC = 5a$. Show that the moment of inertia of this composite body about an axis through A perpendicular to AB and in the plane of the disc is $\frac{77ma^2}{4}$.

The body is held at rest with the end A smoothly hinged to a fixed pivot and with the plane of the disc *horizontal*. The body is released and has angular speed ω when AC is vertical. Find ω in terms of a and g.

When AC is vertical, the centre of the disc strikes a stationary particle of mass $\frac{1}{2}m$. Given that the particle adheres to the centre of the disc, show that the angular speed of the body immediately after impact is $\frac{77\omega}{109}$.

20 Show, by integration, that the moment of inertia of a uniform circular disc, of mass M and radius a, about an axis which passes through its centre and is perpendicular to its plane is $\frac{Ma^2}{2}$.

Without further integration, deduce the moment of inertia of the disc

(a) about an axis perpendicular to its plane and passing through a point on its circumference,

(b) about a diameter.

A uniform disc, of mass M and radius a, is suspended from a smooth pivot at a point on its circumference, and rests in equilibrium. Calculate the period of small oscillations when the centre of the disc is slightly displaced

(c) in the plane of the disc,

(d) perpendicular to the plane of the disc. [E]

21 A compound pendulum consists of a thin uniform rod AB, of length $2a$ and mass $3m$, with a particle of mass $2m$ attached at B. The pendulum is free to rotate in a vertical plane about a horizontal axis perpendicular to the rod through a point C of the rod at a distance $x(<a)$ from A. Show that the moment of inertia of the pendulum about this axis through C is

$$(5x^2 - 14ax + 12a^2)m$$

Find the square of the period of small oscillations of the pendulum. Show that, as x varies, the period takes its minimum value when $x = \frac{[7 - \sqrt{(11)}]a}{5}$. [E]

22 Use integration to show that the moment of inertia of a uniform sphere, of mass m and radius a, about a diameter is $\frac{2}{5}ma^2$. Hence, or otherwise, find the moment of inertia of the sphere about a tangent.

The sphere is free to rotate about a fixed horizontal tangent line with its centre O moving in a vertical plane perpendicular to the tangent. Show that, when the sphere oscillates about the position of stable equilibrium, its equation of motion is the same as that of a simple pendulum of length $\frac{7a}{5}$.

Given that the sphere oscillates between the positions in which O is on the same horizontal level as the tangential axis, find the maximum speed of O during the motion. [E]

23 Show, by integration, that the moment of inertia of a thin uniform circular disc, of radius a and mass m, about an axis through the centre of the disc perpendicular to the plane of the disc is $\frac{1}{2}ma^2$.
Hence determine the moment of inertia of the disc about a parallel axis through a point on the rim of the disc.
The disc is free to rotate in a fixed vertical plane about a fixed horizontal axis through a point on the rim of the disc and at right-angles to the plane of the disc. The diameter through the axis makes an angle θ with the downward vertical. Given that the disc is released from rest when $\theta = \frac{\pi}{2}$, find the speed of the centre of the disc as the centre passes through its lowest point.
In another situation when the disc is at rest with its centre vertically below the axis, it is given a small horizontal impulse acting in the plane of the disc. Show that the motion is approximately simple harmonic, and find the period of the motion. [E]

24 A uniform rod AB, of length $2a$ and mass $6m$, has a particle of mass $2m$ attached at B. The rod is free to rotate in a vertical plane about a smooth fixed horizontal axis perpendicular to the rod and passing through a point X of the rod so that $AX = x$, where $x < a$. Show that the moment of inertia of the system about this axis is

$$4m(4a^2 - 5ax + 2x^2)$$

Find the period of small oscillations of the system about its equilibrium position with B below A. [E]

25 Prove, by integration, that the moment of inertia of a uniform rod of mass m and length a about an axis through its mid-point and perpendicular to the rod is $\frac{1}{12}ma^2$.
Four uniform rods AB, BC, CD and DA, each of length a, are rigidly joined to form a square $ABCD$. Each of the rods AB, CD and DA has mass m and rod BC has mass $3m$. The rods are free to rotate about a smooth horizontal axis, L, which passes through A and is perpendicular to the plane of the square.

(a) Show that the moment of inertia of the system about L is $6ma^2$, and find the distance of the centre of mass of the system from A.

The system is released from rest with AB horizontal and C vertically below B.

(b) Find the greatest value of the angular speed of the system in the subsequent motion.

(c) Find the period of small oscillations of the system about the position of stable equilibrium. [E]

26 Show that the moment of inertia of a uniform solid right circular cone, of mass m, height h and base radius a, about a line through its vertex and perpendicular to its axis of symmetry is

$$\frac{3m(a^2 + 4h^2)}{20}$$

$\left(\text{You may assume that the moment of inertia of a uniform circular disc, of mass } M \text{ and radius } R\text{, about a diameter is } \frac{MR^2}{4}.\right)$

A cone, with $h = \frac{2a}{3}$, is free to rotate about a smooth horizontal axis through its vertex. Find the period of small oscillations under gravity about the stable position of equilibrium. [E]

27 A uniform lamina, of mass m, has the form of a quadrant of a circle of radius a. Show, by integration, that the moment of inertia of the lamina, about an axis l perpendicular to the plane of the lamina and through the centre of the circle of which it is part, is $\frac{1}{2}ma^2$.

The lamina is free to rotate about the axis l, which is horizontal, and when the centre of mass of the lamina is vertically below the axis of rotation the angular speed is Ω. Determine whether the lamina makes complete revolutions in the cases

(a) $\Omega = 2\sqrt{\left(\frac{g}{a}\right)}$,

(b) $\Omega = 3\sqrt{\left(\frac{g}{a}\right)}$.

If the lamina is given a small displacement from its position of stable equilibrium, find the period of its motion as a compound pendulum.

(You may assume that the centre of mass of the lamina is at a distance $\frac{4\sqrt{2}a}{3\pi}$ from the axis of rotation.) [E]

28 Prove that the moment of inertia of a uniform solid sphere of mass M and radius a about a diameter is $\frac{2Ma^2}{5}$. Deduce that the radius of gyration of a uniform solid hemisphere of radius a about any axis through the centre of its plane face is $a\sqrt{\frac{2}{5}}$. Two identical uniform solid hemispheres are such that one can rotate freely about its axis of symmetry, which is fixed, and the other can rotate freely about a fixed axis which coincides with a tangent to the circular rim of its plane face. Find the ratio of their angular speeds when their kinetic energies are equal. [E]

29

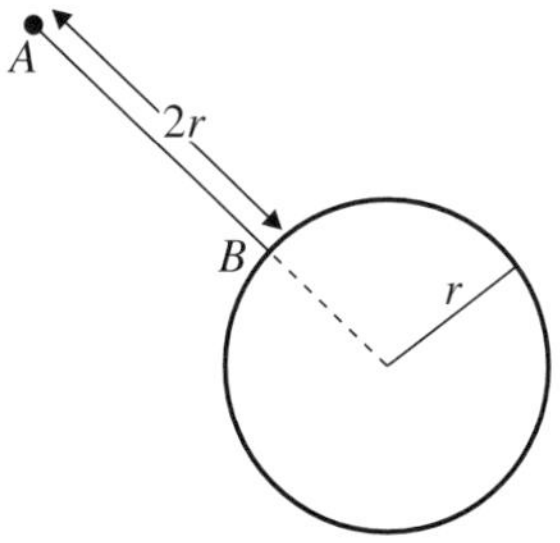

The figure shows a pendulum which can rotate in a vertical plane about a fixed smooth horizontal axis through A. The pendulum consists of a thin uniform rod AB, of mass $\frac{3}{2}m$ and length $2r$, attached to the point B on the surface of a uniform solid sphere, of mass $5m$ and radius r, so that AB produced passes through the centre of the sphere.

(a) Show that the moment of inertia of the pendulum about the horizontal axis through A is $49mr^2$.

(b) Find, in terms of r and g, the period of small oscillations of the pendulum about its position of stable equilibrium. [E]

30 A thin uniform rod AB, of mass M and length $2L$, is freely pivoted at A. The rod hangs vertically with B below A. A particle of mass $\frac{1}{2}M$, travelling horizontally with speed u, strikes the rod at B. After this impact the particle is at rest and the rod starts to move with angular speed ω.

(a) Show that $\omega = \frac{3u}{4L}$.

The rod comes to instantaneous rest when AB is inclined at an angle arccos $(\frac{1}{3})$ to the downward vertical.

(b) Find u in terms of L and g. [E]

31 A uniform circular hoop has mass M and radius a. The points A, B, C and D on the hoop are such that AC and DB are perpendicular diameters of the hoop. Two particles, each of mass m, are attached to the hoop, one at the point A and the other at the point B.

The hoop is free to rotate in a vertical plane about a horizontal axis which passes through D and is perpendicular to the plane of the hoop.

(a) Show that the moment of inertia of the system about the axis is

$$2a^2(M + 3m)$$

The system is released from rest with DB horizontal and A above C.

(b) Find the angular speed ω of the system when DB is vertical.

(c) Show that, if the mass of the hoop is negligible compared with the mass of the particles,

$$\omega^2 = \frac{4g}{3a}$$ [E]

32 Use integration to show that the moment of inertia of a thin uniform rod, of length $2L$ and mass m, about an axis through one end and perpendicular to its length is $\frac{4}{3}mL^2$.

Hence, or otherwise, find the moment of inertia of a uniform square lamina of side $2L$ and mass M about a diagonal. [E]

33 A uniform circular disc of mass m and radius r is free to rotate about a fixed smooth horizontal axis perpendicular to the plane of the disc and at a distance $\frac{1}{2}r$ from the centre of the disc. The disc is held at rest with the centre of the disc vertically above the axis. Given that the disc is slightly disturbed from its position of rest, find the magnitude of the force on the axis when the centre of the disc is in the horizontal plane of the axis. [E]

34 A uniform rod AB, of length $2a$ and mass $2M$, has a particle of mass M attached at the end B. The system is free to rotate in a vertical plane about a fixed horizontal axis through the end A.

(a) Find the moment of inertia of the system about the axis through A.

(b) Show that, if the system is slightly displaced from its position of stable equilibrium it will perform approximate simple harmonic motion with period $2\pi\sqrt{\left(\frac{5a}{3g}\right)}$. [E]

35 A girl sitting on a swing is pulled aside and released. State clearly the assumptions you are making if you consider her motion to be that of a simple pendulum. Given that these assumptions are valid and that the ropes of the swing are of length 2.25 m, calculate the period of oscillation.

Examination style paper

M5

Answer all questions **Time allowed 90 minutes**

Whenever a numerical value of g is required, take $g = 9.8\,\text{m}\,\text{s}^{-2}$.

1. The forces $\mathbf{F}_1 = (b\mathbf{j} + c\mathbf{k})\,\text{N}$, $\mathbf{F}_2 = (c\mathbf{k} + a\mathbf{i})\,\text{N}$ and $\mathbf{F}_3 = (a\mathbf{i} + b\mathbf{j})\,\text{N}$ act through three points with position vectors $a\mathbf{i}\,\text{m}$, $b\mathbf{j}\,\text{m}$ and $c\mathbf{k}\,\text{m}$, respectively, relative to a fixed origin O. Given that a, b and c are constants, show that the force system is equivalent to a single force $\mathbf{F}$ and find the magnitude of $\mathbf{F}$. **(7 marks)**

2. A spherical raindrop falls under gravity through a stationary cloud. Initially the drop is at rest and has radius a. Water from the cloud condenses on the drop as it falls, so that the radius of the drop increases at a constant rate λ. Given that at time t the speed of the drop is v, show that

$$(a + \lambda t)\frac{\mathrm{d}v}{\mathrm{d}t} + 3\lambda v = g(a + \lambda t).$$ **(7 marks)**

3. A simple pendulum, in a vacuum, performs small oscillations of period $\frac{\pi}{2}$ seconds, about its equilibrium position. The pendulum is placed in a fluid and now performs small oscillations under gravity in the fluid which offers a resistance to the motion of the bob. The magnitude of the resistance is $6mv$, where m is the mass and v the speed of the bob.

(*a*) Show that θ, the angular displacement of the pendulum from the vertical, satisfies the differential equation

$$\frac{\mathrm{d}^2\theta}{\mathrm{d}t^2} + 6\frac{\mathrm{d}\theta}{\mathrm{d}t} + 16\theta = 0.$$ **(5 marks)**

(*b*) Show that $\theta = a\mathrm{e}^{-3t}\sin(\sqrt{7}t)$, where a is a small arbitrary constant, gives a possible solution. **(4 marks)**

(*c*) Sketch a graph of θ against t. **(2 marks)**

4. A solid uniform cylinder has mass M, base radius a and height h.

(*a*) Show that the moment of inertia of the cylinder about a diameter of its base is

$$M\left(\frac{a^2}{4} + \frac{h^2}{3}\right).$$ **(9 marks)**

(*b*) Hence obtain the moment of inertia of the cylinder about a parallel axis through its centre of mass. **(3 marks)**

5. A pulley of mass M and radius a is free to rotate in a vertical plane about a smooth, horizontal axis through its centre O. One end of a light, inextensible string is attached to a point on the rim of the pulley. The string is wrapped several times around the pulley and a particle P of mass m is attached to the free end of the string. The system is released from rest.

(a) State a suitable model for the pulley. **(1 mark)**

(b) (i) Show that when P has descended through a distance h its speed is

$$2\left(\frac{mgh}{M+2m}\right)^{\frac{1}{2}}.$$

(ii) Find the tension in the string. **(11 marks)**

6. A uniform square lamina has mass $2m$ and side $2a$. It is smoothly hinged along a horizontal side and hangs vertically. A particle of mass m, moving horizontally with speed v, strikes the lamina perpendicularly at the mid-point of the lower horizontal side and adheres to the lamina. The system starts to rotate about the hinge with an angular speed ω.

(a) Show that $\omega = \dfrac{3v}{10a}$ **(6 marks)**

Given that the system makes complete revolutions,

(b) obtain the condition satisfied by v^2. **(7 marks)**

7. A particle P is moving in a vertical plane. Its velocity $\mathbf{v}$ at time t satisfies the differential equation

$$\frac{d\mathbf{v}}{dt} = -k\mathbf{v} - g\mathbf{j},$$

where k and g are positive constants and $\mathbf{i}$ and $\mathbf{j}$ are unit vectors horizontally and vertically upwards.

Given that $\mathbf{v} = \mathbf{V}$ when $t = 0$,

(a) show that

$$\mathbf{v} = \frac{-g}{k}\mathbf{j} + \left(\mathbf{V} + \frac{g}{k}\mathbf{j}\right)e^{-kt}.$$ **(5 marks)**

(b) What can you deduce about the velocity of P as t becomes very large? **(2 marks)**

Given further that $\mathbf{r} = \mathbf{0}$ when $t = 0$,

(c) obtain an expression for $\mathbf{r}$ as a function of t. **(4 marks)**

(d) If $\mathbf{r} = x\mathbf{i} + y\mathbf{j}$ what can you deduce concerning x from your expression?. **(2 marks)**

Answers

Edexcel accepts no responsibility whatsoever for the accuracy or method of working in the answers given for examination questions.

Exercise 1A

1 $\mathbf{r} = \frac{1}{4}(2\mathbf{i} - 3\mathbf{j})e^{4t} + \frac{7}{2}\mathbf{i} + \frac{3}{4}\mathbf{j}$

2 (a) $\mathbf{r} = -\frac{1}{2}\mathbf{i}e^{-4t} + \frac{7}{2}\mathbf{i} - 3\mathbf{j}$

(b) 4.61 m

3 (a) $\mathbf{v} = (4\mathbf{i} + 2\mathbf{j})e^{2t}$ (b) $13\,300\ \text{m s}^{-1}$

4 $\mathbf{r} = e^{-t}[(\mathbf{i} + \mathbf{j})\cos 2t + (\mathbf{i} + \frac{1}{2}\mathbf{j})\sin 2t]$

0.510 m

5 $\mathbf{r} = (2\mathbf{i} + 2\mathbf{j})\cos 4t + (\mathbf{i} - 2\mathbf{j})\sin 4t$

6 $\mathbf{r} = e^{-t}[(2\mathbf{i} + \mathbf{j})t + \mathbf{i} + 2\mathbf{j}]$

7 $\mathbf{r} = (3\mathbf{i} + 2\mathbf{j})e^{3t} - (\mathbf{i} + \mathbf{j})^{-2t}$

8 $\mathbf{r} = \frac{1}{2}e^{2t}(\mathbf{i} + \mathbf{j} + \mathbf{k}) + \frac{3}{2}\mathbf{i} - \frac{1}{2}\mathbf{j} + \frac{7}{2}\mathbf{k}$

$94.6\ \text{m s}^{-1}$

9 (a) $\mathbf{r} = e^{-2t}[(\cos t + 2\sin t)\mathbf{i}$
$+(2\cos t + 4\sin t)\mathbf{j} + (\cos t + 3\sin t)\mathbf{k}]$

(b) 0.233 m

10 $\mathbf{r} = \frac{1}{2}\mathbf{k}e^{t} + (2\mathbf{i} + 3\mathbf{j} + \frac{1}{2}\mathbf{k})e^{-t}$

11 $\mathbf{r} = e^{-t}[(\mathbf{i} + \mathbf{j})\cos 2t + \mathbf{i}\sin 2t]$, $t = \frac{1}{2}\arctan\frac{1}{3}$

12 $\mathbf{r} = 2\mathbf{i} + \frac{2}{3}\mathbf{j} + \frac{7}{3}\mathbf{k} - (\mathbf{i} + \frac{2}{3}\mathbf{j} + \frac{1}{3}\mathbf{k})e^{-3t}$

13 $\mathbf{r} = [(\mathbf{i} + 2\mathbf{j} + 2\mathbf{k})t - \mathbf{k}]e^{2t} + 2\mathbf{i}$

14 $\mathbf{r} = \frac{1}{4}(5\mathbf{i} + \mathbf{j} + 2\mathbf{k})e^{2t} + \frac{1}{4}(-\mathbf{i} - 3\mathbf{j} + 2\mathbf{k})e^{-2t}$
$-3t\mathbf{i} + \frac{1}{2}\mathbf{j}$

15 $\mathbf{r} = 2e^{-t}\mathbf{j} + (2\mathbf{i} - 3\mathbf{j})e^{-3t}$

16 $\mathbf{r} = \frac{1}{6}(7\mathbf{i} + 11\mathbf{j})e^{4t} + \frac{1}{6}(5\mathbf{i} + 13\mathbf{j})e^{-2t} - (\mathbf{i} + 2\mathbf{j})e^{t}$

Exercise 1B

1 (a) 12 J (b) 19 J (c) 3 J

2 (a) 23 J (b) 24 J (c) 8 J

3 (a) $4.29\ \text{m s}^{-1}$

(b) $4.38\ \text{m s}^{-1}$

(c) $2.53\ \text{m s}^{-1}$

4 (a) $\frac{308}{45}\mathbf{i} + \frac{619}{45}\mathbf{j} + \frac{70}{9}\mathbf{k}$

(b) 20.4 J

5 (a) $(3\mathbf{i} + 5\mathbf{j} + 2\mathbf{k})\ \text{m s}^{-2}$

(b) $(\frac{3}{2}\mathbf{i} + \frac{5}{2}\mathbf{j} + \mathbf{k})$ N

(c) 38 J

6 23 J

7 (a) $(-26\mathbf{i} - 4\mathbf{j} + 22\mathbf{k})$ N m

(b) $(39\mathbf{i} + 3\mathbf{j} + 69\mathbf{k})$ N m

(c) $(-13\mathbf{i} + 10\mathbf{j} - 24\mathbf{k})$ N m

8 (a) $(-23\mathbf{i} - 10\mathbf{j} + 25\mathbf{k})$ N m

(b) $(-13\mathbf{i} + 10\mathbf{j} + 24\mathbf{k})$ N m

(c) $(-19\mathbf{i} - 21\mathbf{j} - 3\mathbf{k})$ N m

9 (a) $(36\mathbf{i} - 72\mathbf{j} - 27\mathbf{k})$ N m

(b) $(-21\mathbf{j})$ N m

10 8, 4, 1

Exercise 1C

1 (a) $(6\mathbf{i} + 4\mathbf{j} - 5\mathbf{k})$ N

(b) $(13\mathbf{i} + 4\mathbf{j} + 2\mathbf{k})$ N

2 (a) $(-4\mathbf{i} - 5\mathbf{j})$ N

(b) $(-9\mathbf{i} - 3\mathbf{j} + 3\mathbf{k})$ N

(c) $(-2\mathbf{i} - 3\mathbf{j} - 7\mathbf{k})$ N

4 (a) $(2\mathbf{i} - \mathbf{j} + 8\mathbf{k})$ N $(8\mathbf{i} - 9\mathbf{j} - 8\mathbf{k})$ N m

(b) $(2\mathbf{i} - \mathbf{j} + 8\mathbf{k})$ N $(14\mathbf{i} - 5\mathbf{j} - 9\mathbf{k})$ N m

(c) $\mathbf{r} = \lambda(2\mathbf{i} - \mathbf{j} + 8\mathbf{k})$

$\mathbf{r} = \mathbf{i} - \mathbf{j} + 2\mathbf{k} + \mu(2\mathbf{i} - \mathbf{j} + 8\mathbf{k})$

5 $(-10\mathbf{j} - 13\mathbf{k})$ N m

6 $(5\mathbf{i} + 22\mathbf{j} + 11\mathbf{k})$ N $(117\mathbf{i} - 16\mathbf{j} - 55\mathbf{k})$ N m

7 $\mathbf{R} = (9\mathbf{i} + 4\mathbf{j} + 6\mathbf{k})$ N

$\mathbf{r} = 7\mathbf{i} + 6\mathbf{j} + \lambda(9\mathbf{i} + 4\mathbf{j} + 6\mathbf{k})$

8 (a) $(-2\mathbf{i} - \mathbf{k})$ N (b) 11.4 N m

9 (a) $\mathbf{F}_3 = (-7\mathbf{i} + \mathbf{j} + 2\mathbf{k})$ N

10 $p = 2$, $q = 1$, $r = -8$, $6\mathbf{i} - 18\mathbf{j}$

11 $\mathbf{r} = \frac{1}{4}(-2\mathbf{i} + 4\mathbf{j} + 3\mathbf{k}) + \lambda(2\mathbf{i} + 2\mathbf{j} - \mathbf{k})$

Exercise 2A

1 If the thrust were not greater than its weight, the rocket would not lift off.

2 $\dfrac{g}{8k^2}(kt+a)^2 + \dfrac{ga^4}{8k^2}\dfrac{1}{(kt+a)^2} - \dfrac{ga^2}{4k^2}$

Review exercise 1

1 (a) $\dfrac{\omega^2}{2}ma^2(9\sin^2\omega t + 16\cos^2\omega t)$

(b) $ma\omega^2(9\cos^2\omega t + 16\sin^2\omega t)^{\frac{1}{2}}$

$\mathbf{r} = 3a\cos\omega t\mathbf{i} + a\sin\omega t\mathbf{j} - 4a\cos\omega t\mathbf{k}$

$\mathbf{r}\cdot\mathbf{r} = a^2(\sin^2\omega t + 25\cos^2\omega t)$

2 $\mathbf{r} = (e^\theta - 1)\mathbf{i} + \mathbf{j}$

3 $\mathbf{r} = e^\theta(A\cos 3\theta + B\sin 3\theta)$

$\mathbf{r} = -e^\pi\mathbf{j}$

4 $\mathbf{r} = a(2\mathbf{i}+\mathbf{j}) - a(\mathbf{i}-\mathbf{j})e^{-\frac{3t}{T}}$

5 $\mathbf{r} = a(\mathbf{i}+\mathbf{j})\cos nt + \dfrac{b}{n}(\mathbf{i}-2\mathbf{k})\sin nt$

6 $\mathbf{r} = (\mathbf{i}-\mathbf{k})e^{2t}$

7 $\mathbf{r} = \left(\dfrac{5a}{13}\sinh 3\omega t - \dfrac{a}{13}\sin 2\omega t\right)\mathbf{i} + (a\cosh 3\omega t)\mathbf{j}$

8 5 J

9 7 J

10 240 J

11 $\dfrac{2\sqrt{53}}{27}(7\mathbf{i}+4\mathbf{j}+4\mathbf{k})$, 52 J

12 $3\mathbf{i}+\mathbf{j}+2\mathbf{k}$, $4\mathbf{i}-4\mathbf{j}-4\mathbf{k}$

13 $(4\mathbf{k}-3\mathbf{j})$ N m, 5 N m

14 $2\sqrt{(a^2+b^2+c^2)}$

15 $(\mathbf{i}+2\mathbf{j}+3\mathbf{k})$ m, $a=7$, $b=2$

16 $(18\mathbf{i}-56\mathbf{j}-20\mathbf{k})$ N m, 160 J,

$(6\mathbf{i}-5\mathbf{j}+23\mathbf{k})$ N, $(16\mathbf{i}-57\mathbf{j}-17\mathbf{k})$ N m

17 (a) 5λ

(b) $-2\lambda a\mathbf{k}$

(c) $4x+3y=2$

18 $(-3\mathbf{i}+4\mathbf{j}-\mathbf{k})$ N, $\mathbf{r} = (3\mathbf{i}-\mathbf{k}) + \lambda(3\mathbf{i}-4\mathbf{j}+\mathbf{k})$,

$(4\mathbf{i}+6\mathbf{j}+12\mathbf{k})$ N m

19 $(3\mathbf{i}+2\mathbf{j}-4\mathbf{k})$ N

20 $\sqrt{2}$ N

21 (a) $\mathbf{F} = (\mathbf{i}-3\mathbf{j}+2\mathbf{k})$ m

(b) 17 N m

22 (a) $h=9$

(b) $\mathbf{G} = (-4\mathbf{i}+17\mathbf{j}+10\mathbf{k})$ N m

23 $\mathbf{F} = 4\mathbf{i}+6\mathbf{j}+4\mathbf{k}$

24 (a) $\frac{1}{3}[(\mathbf{i}+2\mathbf{j}) + (2\mathbf{i}+\mathbf{j})e^{6t}]$ m

(b) 0.22 s

25 $\mathbf{r} = (\mathbf{i}+5\mathbf{j})e^{-t} + (-\mathbf{i}-3\mathbf{j})e^{-2t}$

26 $\mathbf{r} = (5.75\mathbf{i}-14)$ m

29 $\frac{3}{2}m$

30 $\dfrac{g}{3\lambda}\left[m+\lambda t - \dfrac{m^3}{(m+\lambda t)^2}\right] + \dfrac{m^2u}{(m+\lambda t)^2}$

31 $\dfrac{1}{k}\ln\cosh\left[t\sqrt{\left(\dfrac{g}{k}\right)}\right]$

32 $v = -u\ln(1-kt) - gt$

33 $\frac{1}{8}Mu^2(1-k)[\ln(1-k)]^2$, $1-e^{-2}$

Exercise 3A

1 $\dfrac{4ma^2}{3}$, $\dfrac{2a\sqrt{3}}{3}$

2 $\dfrac{m}{3}(l^2-3la+3a^2)$, $\sqrt{\left(\dfrac{l^2-3la+3a^2}{3}\right)}$

3 $\frac{1}{3}ml^2\sin^2\theta$, $\dfrac{\sqrt{3}}{3}l\sin\theta$

4 $\frac{1}{6}mb^2$, $\frac{1}{6}b\sqrt{6}$

5 $\frac{16}{5}ma^2$, $\dfrac{4a\sqrt{5}}{5}$

6 $7mr^2$

7 $\frac{1}{3}ml^2$

8 $\frac{1}{12}ml^2$

9 $\frac{11}{9}ma^2$

10 $\frac{41}{72}md^2$

11 $\frac{5}{2}ml^2$

12 $\dfrac{Mr^2(2h+r)}{2(h+r)}$

13 $\frac{1}{3}m(a^2+b^2)$

14 $\frac{1}{2}Mr^2$

15 $\pi R^3\rho(R+4H)$

16 mr^2

17 $24ma^2$

18 $5mr^2$

Exercise 3B

1 $\frac{3}{10}mr^2$

2 $\frac{3}{2}mr^2$

3 $\frac{1}{4}mr^2$, $\frac{5}{4}mr^2$

4 md^2

5 (a) $\frac{16}{3}ma^2$ (b) $\frac{40}{3}ma^2$ (c) $\frac{28}{3}ma^2$

6 (a) $\frac{29}{3}mr^2$ (b) $\frac{41}{3}mr^2$

(c) $\frac{79}{6}mr^2$ (d) $\frac{173}{18}mr^2$

7 $\frac{1}{12}m(a^2+4b^2)$

8 $r\sqrt{(\frac{3}{2})}$

10 $\dfrac{8\sqrt{3}}{3}$

11 $\dfrac{7mr^2}{5}$

12 $\frac{1}{4}\sqrt{6}$

13 (a) $\frac{5}{2}ma^2$ (b) $\frac{13}{2}ma^2$

14 $\frac{1}{3}a\sqrt{3}$

15 $8Ml^2$

16 $\dfrac{5mr^2}{2}$

17 $\dfrac{15mr^2}{2}$

18 $\frac{2}{5}mr^2$

19 $\dfrac{m}{10}(9r^2+46h^2)$

20 (b) $\dfrac{Ma^2(16+15\pi)}{4(4+\pi)}$

Exercise 4A

1 8 J

2 1.58 J, 1.58 J

3 (a) 4.35 J (b) 4.08 J

4 (a) 14.7 J

(b) 4.43 rad s^{-1}

(c) 6.26 rad s^{-1}

5 2.56 rad s^{-1}, particle

6 (a) $\sqrt{\left(\dfrac{6g}{13l}\right)}$ (b) 1.65^c

7 0.506^c

8 (a) $4\sqrt{\left(\dfrac{3g}{19l}\right)}$

(b) $20\sqrt{\left(\dfrac{gl}{19}\right)}$

(c) $\frac{\pi}{3}$

9 (a) $\dfrac{9l}{8}$ (b) $\frac{243}{128}mgl$

10 $\sqrt{\left(\dfrac{g}{a}\right)}$ rad s^{-1}

11 $\sqrt{\left(\dfrac{2mgh}{M+m}\right)}$

12 $k>3$, uniform lamina

13 $\sqrt{\dfrac{6g}{11a}}$

14 $\dfrac{3}{2}\sqrt{\dfrac{g}{2r}}$

15 $2\sqrt{\dfrac{15g}{31r}}$

16 $\sqrt{\left(\dfrac{20gh}{(3r^2+2h^2)}\right)}$

Exercise 4B

1 $\dfrac{15g}{2a}$

2 0.25 N

3 5 N, 1.04 kg

4 (a) $\dfrac{2g}{5a}$ (b) $\dfrac{2g}{5a}\cos\theta$

5 3.06 N, 3.68 N, 4.9 rad s^{-2}

6 (a) $\dfrac{13mg}{7}$ (b) $\dfrac{\sqrt{217}}{14}mg$

7 (a) $\sqrt{\left(\dfrac{3g(1-\cos\theta)}{7a}\right)}$

(b) $\dfrac{19mg}{7}$

8 (a) $mg\left(4\sqrt{2}-\dfrac{3}{2}\right)$, $\dfrac{mg}{4}$

(b) 0, $4mg(\sqrt{2}-1)$

9 (a) $\dfrac{2mg}{5}$, $\dfrac{14mg}{15}$ (b) 0, $\dfrac{19mg}{5}$

10 (a) $\sqrt{\dfrac{3g}{2l}}$ (b) $\dfrac{7mg}{4}$

11 (a) $\sqrt{\dfrac{66g}{67l}}$ (b) $\dfrac{1128mg}{67}$

12 (a) $\dfrac{36mg}{13}$, $\dfrac{34mg}{13}$ (b) $\dfrac{4g}{13}$

13 $mg\sqrt{111}$

Exercise 4C

1 15 N m s

2 $\frac{1}{2}$ N m s

3 $\dfrac{7}{34}\sqrt{\left(\dfrac{6g}{7a}\right)}$ rad s^{-1} 0.322^c

4 5.42 rad s^{-1}, 7.23 N s

5 $2m\sqrt{\left(\frac{ga}{3}\right)}$, $\frac{\pi}{3}$

6 0.537 m

7 2.86 rad s^{-1}

8 $a(4 - \sqrt{2})$

9 $\frac{27u}{8a}$

10 $\frac{18v}{13l}$

11 $68.1m\sqrt{(gl)}$, yes

12 $\frac{\omega}{10}$

13 $m\sqrt{\frac{14ag}{27}}$

14 $\frac{4}{3}\sqrt{\frac{2g}{r}}$

15 $\frac{4}{3}\sqrt{\frac{g}{a}}$

Exercise 4D

1 (a) 1.42 s (b) 2.29 s (c) 1.68 s

2 (a) 0.345 m (b) 0.484 m (c) 0.248 m

3 0.993 m

4 $\frac{T}{\sqrt{2}}$

5 $9L$

6 Cord is inelastic; metal bob is small and heavy compared with cord; amplitude of oscillations is small.
2.84 s

7 Rope is long compared with height of boy; rope is light compared with weight of boy; rope is inelastic.
3.04 m

8 (i) $4\pi\sqrt{\left(\frac{a}{3g}\right)}$ (ii) $\frac{4a}{3}$

9 (i) $4\pi\sqrt{\left(\frac{10a}{21g}\right)}$ (ii) $\frac{40a}{21}$

10 (i) $2\pi\sqrt{\left(\frac{d}{g}\right)}$ (ii) d

11 (i) $2\pi\sqrt{\left(\frac{a\sqrt{3}}{g}\right)}$ (ii) $a\sqrt{3}$

12 (i) 3.28 s (ii) $2\frac{2}{3}$ m

13 (i) $2\pi\sqrt{\left(\frac{3r}{2g}\right)}$ (ii) $\frac{3r}{2}$

14 $4\pi\sqrt{\left(\frac{a}{3g}\right)}$, $\frac{11m}{3}$

15 (a) $2\pi\sqrt{\left(\frac{Ma^2 + 3mx^2}{3mgx}\right)}$

(c) $a\sqrt{\left(\frac{M}{3m}\right)}$

16 (b) $6a$

17 $\frac{146}{15}ma^2$, $2.99a$

18 (b) $\frac{4\pi}{3}\sqrt{\frac{7a}{g}}$

20 $2\pi\sqrt{\frac{287l}{80g}}$

21 $2\pi\sqrt{\frac{23b}{11g}}$

Review exercise 2

1 $\frac{1}{4}ma^2$, $\frac{5}{4}ma^2$

2 $\frac{2}{3}Ma^2$

3 (a) $\frac{7}{2}Ma^2$ (b) $\frac{9}{4}Ma^2$

4 (a) $\sqrt{\frac{13}{8}}$ (b) $\frac{7\sqrt{2}}{8}$

6 $\frac{2V^2}{3g}$, $\frac{3g}{4}$, $\frac{3mg}{4}$

7 109 rpm

8 2.03 kg m^2

9 $\left(\frac{2\sqrt{3g}}{3l}\right)^{\frac{1}{2}}$

10 $\left(\frac{2mgx}{M+m}\right)^{\frac{1}{2}}$

11 $\left(\frac{6g(\sqrt{2}-1)}{5a}\right)^{\frac{1}{2}}$, $(2.2 - 0.6\sqrt{2})Mg$

12 $\frac{8}{3}ma^2$, $\frac{5}{2}mg\cos\theta + m\sqrt{2}a\omega^2 - \frac{3}{2}mg$
$\frac{1}{4}mg\sin\theta$

13 $\frac{5}{2}Mg\sin\theta$, $\frac{1}{4}Mg\cos\theta$

14 $\frac{4}{7}mg\cos\theta$, $\frac{13}{7}mg\sin\theta$

16 $\frac{3v}{4a}$

17 $\frac{Ma^2}{2}$, $\left(\frac{8g}{5a}\right)^{\frac{1}{2}}$, $\frac{4u}{9a} + \frac{5}{9}\left(\frac{8g}{5a}\right)^{\frac{1}{2}}$

18 $\dfrac{3Vd}{3d^2 + 8a^2}$

19 $\sqrt{\left(\dfrac{4g}{7a}\right)}$

20 (a) $\dfrac{3Ma^2}{2}$ (b) $\dfrac{Ma^2}{4}$

(c) $\pi\sqrt{\left(\dfrac{6a}{g}\right)}$ (d) $\pi\sqrt{\left(\dfrac{5a}{g}\right)}$

21 $\dfrac{4\pi^2(12a^2 - 14ax + 5x^2)}{(7a - 5x)g}$

22 $\dfrac{7ma^2}{5}$, $\sqrt{\left(\dfrac{10ga}{7}\right)}$

23 $\frac{3}{2}ma^2$, $2\sqrt{\left(\dfrac{ga}{3}\right)}$, $2\pi\sqrt{\left(\dfrac{3a}{2g}\right)}$

24 $2\pi\left(\dfrac{2(4a^2 - 5ax + 2x^2)}{g(5a - 4x)}\right)^{\frac{1}{2}}$

25 (a) $\dfrac{5a}{6}$ (b) $\sqrt{\left(\dfrac{2g}{3a}\right)}$

(c) $2\pi\sqrt{\left(\dfrac{6a}{5g}\right)}$

26 $2\pi\sqrt{\left(\dfrac{5a}{6g}\right)}$

27 (a) no

(b) yes; $\pi\sqrt{\left(\dfrac{3\pi a}{2\sqrt{2g}}\right)}$

28 $\sqrt{7} : \sqrt{2}$

29 (b) $2\pi\sqrt{\dfrac{98r}{33g}}$

30 (b) $\frac{4}{3}\sqrt{(gl)}$

31 (b) $\sqrt{\dfrac{g(M + 4m)}{a(M + 3m)}}$

32 $\frac{1}{3}ML^2$

33 $\dfrac{2\sqrt{2}}{3}mg$

34 (a) $\dfrac{20Ma^2}{3}$

35 No friction, light inextensible chain, model girl and seat as point mass.

3.01s

Examination style paper

1 $|\mathbf{F}| = 2(a^2 + b^2 + c^2)^{\frac{1}{2}}$

3 (c)

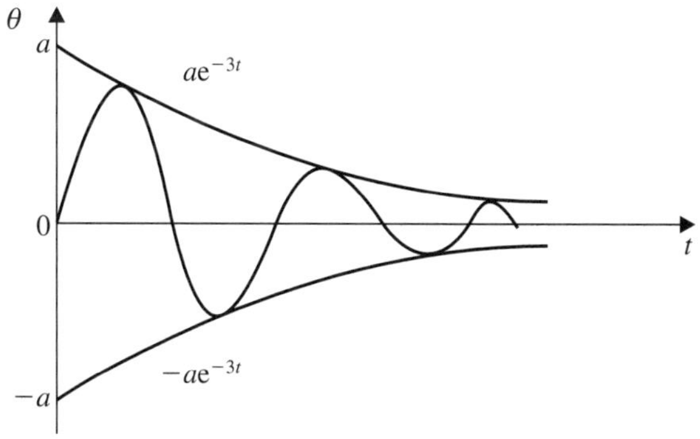

4 (b) $M\left(\dfrac{a^2}{4} + \dfrac{h^2}{12}\right)$

5 (b) (ii) $T = \dfrac{Mmg}{(M + 2m)}$

6 (b) $v^2 > \dfrac{80}{3}ag$

7 (b) As $t \to \infty$, $\mathbf{v} \to \dfrac{-g}{k}\mathbf{j}$

Velocity tends to vertically downwards.

Terminal speed $\dfrac{g}{k}$.

(c) $\mathbf{r} = \dfrac{-g}{k}t\mathbf{j} - \dfrac{1}{k}\left(\mathbf{V} + \dfrac{g}{k}\mathbf{j}\right)(e^{-kt} - 1)$

(d) As $t \to \infty$, $x \to \dfrac{V_1}{k}$

where $\mathbf{V} = V_1\mathbf{i} + V_2\mathbf{j}$

List of symbols and notation

The following notation will be used in all Edexcel examinations.

$\in$	is an element of
$\notin$	is not an element of
$\{x_1, x_2, \ldots\}$	the set with elements $x_1, x_2, \ldots$
$\{x : \ldots\}$	the set of all x such that ...
$\mathrm{n}(A)$	the number of elements in set A
$\varnothing$	the empty set
$\mathscr{E}$	the universal set
A	the complement of the set A
$\mathbb{N}$	the set of natural numbers, $\{1, 2, 3, \ldots\}$
$\mathbb{Z}$	the set of integers, $\{0, \pm 1, \pm 2, \pm 3, \ldots\}$
$\mathbb{Z}^+$	the set of positive integers, $\{1, 2, 3, \ldots\}$
$\mathbb{Z}_n$	the set of integers modulo n, $\{0, 1, 2, \ldots, n-1\}$
$\mathbb{Q}$	the set of rational numbers $\left\{\frac{p}{q} : p \in \mathbb{Z}, q \in \mathbb{Z}^+\right\}$
$\mathbb{Q}^+$	the set of positive rational numbers, $\{x \in \mathbb{Q} : x > 0\}$
$\mathbb{Q}_0^+$	the set of positive rational numbers and zero, $\{x \in \mathbb{Q} : x \geqslant 0\}$
$\mathbb{R}$	the set of real numbers
$\mathbb{R}^+$	the set of positive real numbers, $\{x \in \mathbb{R} : x > 0\}$
$\mathbb{R}_0^+$	the set of positive real numbers and zero, $\{x \in \mathbb{R} : x \geqslant 0\}$
$\mathbb{C}$	the set of complex numbers
(x, y)	the ordered pair x, y
$A \times B$	the cartesian product of sets A and B, $A \times B = \{(a, b) : a \in A, b \in B\}$
$\subseteq$	is a subset of
$\subset$	is a proper subset of
$\cup$	union
$\cap$	intersection
$[a, b]$	the closed interval, $\{x \in \mathbb{R} : a \leqslant x \leqslant b\}$
$[a, b)$	the interval $\{x \in \mathbb{R} : a \leqslant x < b\}$
$(a, b]$	the interval $\{x \in \mathbb{R} : a < x \leqslant b\}$
(a, b)	the open interval $\{x \in \mathbb{R} : a < x < b\}$
$y R x$	y is related to x by the relation R
$y \sim x$	y is equivalent to x, in the context of some equivalence relation
$=$	is equal to
$\neq$	is not equal to

$\equiv$	is identical to *or* is congruent to
$\approx$	is approximately equal to
$\cong$	is isomorphic to
$\propto$	is proportional to
$<$	is less than
$\leqslant, \ngtr$	is less than or equal to, is not greater than
$>$	is greater than
$\geqslant, \nless$	is greater than or equal to, is not less than
∞	infinity
$p \wedge q$	p and q
$p \vee q$	p or q (or both)
$\sim p$	not p
$p \Rightarrow q$	p implies q (if p then q)
$p \Leftarrow q$	p is implied by q (if q then p)
$p \Leftrightarrow q$	p implies and is implied by q (p is equivalent to q)
$\exists$	there exists
$\forall$	for all
$a + b$	a plus b
$a - b$	a minus b
$a \times b$, ab, $a.b$	a multiplied by b
$a \div b$, $\frac{a}{b}$, a/b	a divided by b
$\sum_{i=1}^{n} a_i$	$a_1 + a_2 + \ldots + a_n$
$\prod_{i=1}^{n} a_i$	$a_1 \times a_2 \times \ldots \times a_n$
$\sqrt{a}$	the positive square root of a
$\lvert a \rvert$	the modulus of a
$n!$	n factorial
$\binom{n}{r}$	the binomial coefficient $\frac{n!}{r!(n-r)!}$ for $n \in \mathbb{Z}^+$; $\frac{n(n-1)\ldots(n-r+1)}{r!}$ for $n \in \mathbb{Q}$
$\mathrm{f}(x)$	the value of the function f at x
$\mathrm{f}: A \to B$	f is a function under which each element of set A has an image in set B
$\mathrm{f}: x \mapsto y$	the function f maps the element x to the element y
f^{-1}	the inverse function of the function f
$\mathrm{g} \circ \mathrm{f}$, gf	the composite function of f and g which is defined by $(\mathrm{g} \circ \mathrm{f})(x)$ or $\mathrm{gf}(x) = \mathrm{g}(\mathrm{f}(x))$
$\lim_{x \to a} \mathrm{f}(x)$	the limit of $\mathrm{f}(x)$ as x tends to a
Δx, δx	an increment of x
$\frac{\mathrm{d}y}{\mathrm{d}x}$	the derivative of y with respect to x
$\frac{\mathrm{d}^n y}{\mathrm{d}x^n}$	the nth derivative of y with respect to x

$\mathrm{f}(x), \mathrm{f}(x), \ldots \mathrm{f}^{(n)}(x)$	the first, second, ... nth derivatives of $\mathrm{f}(x)$ with respect to x
$\int y \, \mathrm{d}x$	the indefinite integral of y with respect to x
$\int_a^b y \, \mathrm{d}x$	the definite integral of y with respect to x between the limits $x = a$ and $x = b$
$\frac{\partial V}{\partial x}$	the partial derivative of V with respect to x
$\dot{x}, \ddot{x}, \ldots$	the first, second, . . . derivatives of x with respect to t
e	base of natural logarithms
e^x, $\exp x$	exponential function of x
$\log_a x$	logarithm to the base a of x
$\ln x$, $\log_\mathrm{e} x$	natural logarithm of x
$\lg x$, $\log_{10} x$	logarithm to the base 10 of x
sin, cos, tan cosec, sec, cot	the circular functions
arcsin, arccos, arctan arccosec, arcsec, arccot	the inverse circular functions
sinh, cosh, tanh cosech, sech, coth	the hyperbolic functions
arsinh, arcosh, artanh, arcosech, arsech, arcoth	the inverse hyperbolic functions
i	square root of -1
z	a complex number, $z = x + \mathrm{i}y$
Re z	the real part of z, Re $z = x$
Im z	the imaginary part of z, Im $z = y$
$\|z\|$	the modulus of z, $\|z\| = \sqrt{(x^2 + y^2)}$
arg z	the argument of z, arg $z = \arctan \frac{y}{x}$
z^*	the complex conjugate of z, $x - \mathrm{i}y$
$\mathbf{M}$	a matrix $\mathbf{M}$
$\mathbf{M}^{-1}$	the inverse of the matrix $\mathbf{M}$
$\mathbf{M}^\mathrm{T}$	the transpose of the matrix $\mathbf{M}$
det $\mathbf{M}$, $\|\mathbf{M}\|$	the determinant of the square matrix $\mathbf{M}$
$\mathbf{a}$	the vector $\mathbf{a}$
$\overrightarrow{AB}$	the vector represented in magnitude and direction by the directed line segment AB
$\hat{\mathbf{a}}$	a unit vector in the direction of $\mathbf{a}$
$\mathbf{i}, \mathbf{j}, \mathbf{k}$	unit vectors in the directions of the cartesian coordinate axes
$\|\mathbf{a}\|$, a	the magnitude of $\mathbf{a}$
$\|\overrightarrow{AB}\|$, AB	the magnitude of $\overrightarrow{AB}$
$\mathbf{a} \,.\, \mathbf{b}$	the scalar product of $\mathbf{a}$ and $\mathbf{b}$
$\mathbf{a} \times \mathbf{b}$	the vector product of $\mathbf{a}$ and $\mathbf{b}$

A, B, C, etc	events
$A \cup B$	union of the events A and B
$A \cap B$	intersection of the events A and B
$\mathrm{P}(A)$	probability of the event A
A	complement of the event A
$\mathrm{P}(A \mid B)$	probability of the event A conditional on the event B
X, Y, R, etc.	random variables
x, y, r, etc.	values of the random variables X, Y, R, etc
$x_1, x_2 \ldots$	observations
$f_1, f_2, \ldots$	frequencies with which the observations $x_1, x_2, \ldots$ occur
$\mathrm{p}(x)$	probability function $\mathrm{P}(X = x)$ of the discrete random variable X
$p_1, p_2, \ldots$	probabilities of the values $x_1, x_2, \ldots$ of the discrete random variable X
$\mathrm{f}(x), \mathrm{g}(x), \ldots$	the value of the probability density function of a continuous random variable X
$\mathrm{F}(x), \mathrm{G}(x), \ldots$	the value of the (cumulative) distribution function $\mathrm{P}(X \leqslant x)$ of a continuous random variable X
$\mathrm{E}(X)$	expectation of the random variable X
$\mathrm{E}[\mathrm{g}(X)]$	expectation of $\mathrm{g}(X)$
$\mathrm{Var}(X)$	variance of the random variable X
$\mathrm{G}(t)$	probability generating function for a random variable which takes the values 0, 1, 2, ...
$\mathrm{B}(n, p)$	binomial distribution with parameters n and p
$\mathrm{N}(\mu, \sigma^2)$	normal distribution with mean μ and variance σ^2
μ	population mean
σ^2	population variance
σ	population standard deviation
$\bar{x}$, m	sample mean
s^2, $\hat{\sigma}^2$	unbiased estimate of population variance from a sample, $s^2 = \frac{1}{n-1}\sum(x_i - \bar{x})^2$
ϕ	probability density function of the standardised normal variable with distribution N(0, 1)
Φ	corresponding cumulative distribution function
ρ	product-moment correlation coefficient for a population
r	product-moment correlation coefficient for a sample
$\mathrm{Cov}\,(X, Y)$	covariance of X and Y

Index